SELECTED TOPICS IN GEOMETRY WITH CLASSICAL VS. COMPUTER PROVING

SELECTED TOPICS IN GEOMETRY WITH CLASSICAL VS. COMPUTER PROVING

PAVEL PECH

University of South Bohemia
Czech Republic

NEW JERSEY • LONDON • SINGAPORE • BEIJING • SHANGHAI • HONG KONG • TAIPEI • CHENNAI

Published by

World Scientific Publishing Co. Pte. Ltd.

5 Toh Tuck Link, Singapore 596224

USA office: 27 Warren Street, Suite 401-402, Hackensack, NJ 07601

UK office: 57 Shelton Street, Covent Garden, London WC2H 9HE

Library of Congress Cataloging-in-Publication Data

Pech, Pavel.

Selected topics in geometry with classical vs. computer proving / Pavel Pech.

p. cm.

Includes bibliographical references and index.

ISBN-13 978-981-270-942-4 (hardcopy : alk. paper)

ISBN-10 981-270-942-8 (hardcopy : alk. paper)

1. Geometry. 2. Geometry--Computer programs. 3. Problem solving. I. Title.

QA445.P42 2007

516--dc22

2007028204

British Library Cataloguing-in-Publication Data

A catalogue record for this book is available from the British Library.

Printed in Singapore.

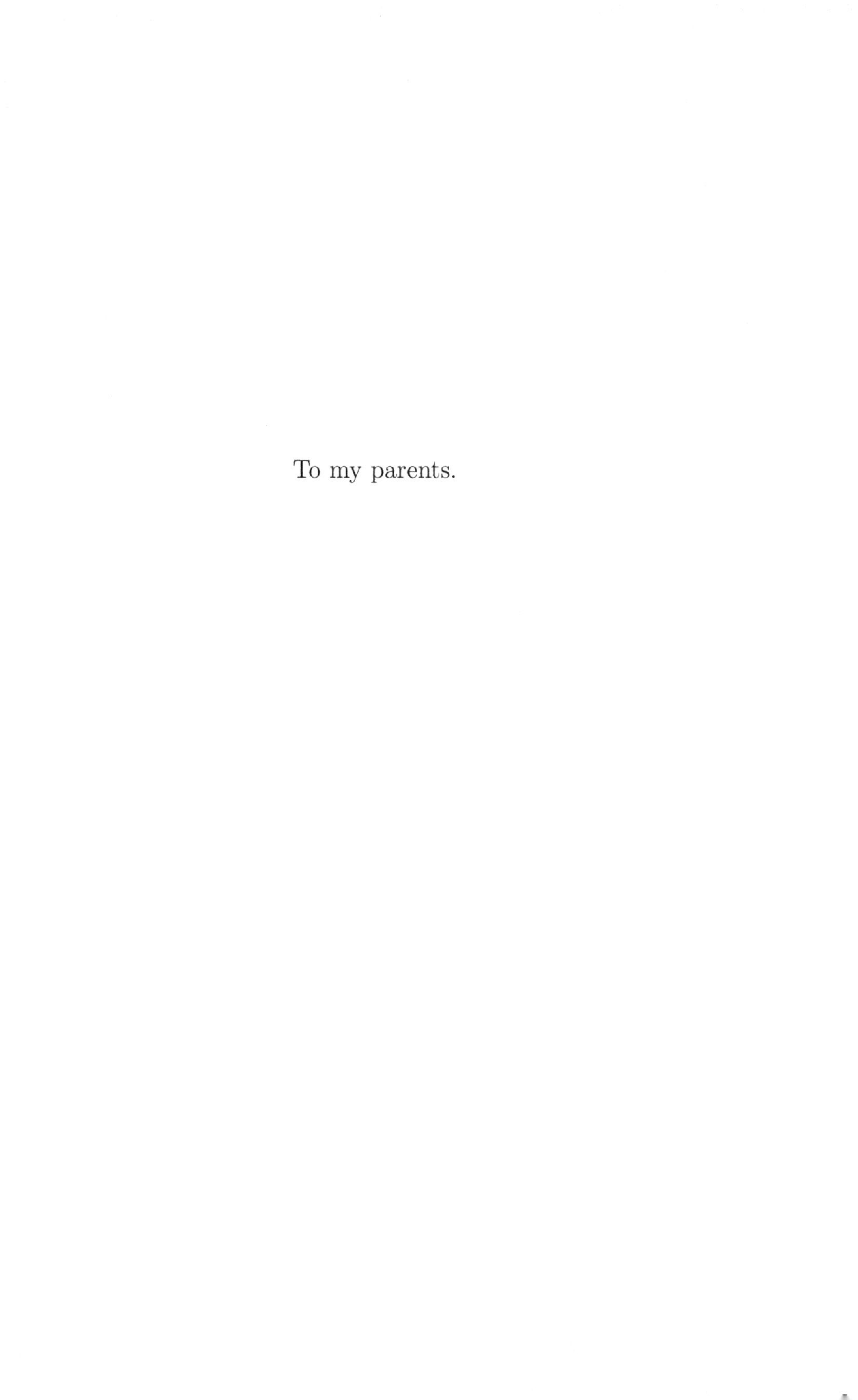

To my parents.

Preface

This book has essentially two objectives. One of them is to give you the basic notions of the theory of automatic theorem proving, which was developed in the last third of the 20th century, and apply this theory to examples. The second objective is to show the interesting topics of elementary geometry which the author met in the course of years.

The content of this book consists of six mutually independent stories, which are aimed at a certain theme. Each of these stories was thoroughly studied by the author in the past and an attempt is now made to put together all these themes. Each problem is firstly solved by the method of automatic theorem proving, whereas in the second step the problem is solved classically – without using computer (of course, if it is possible).

The author's aim is to acquaint the reader with new computer methods of proving, deriving and discovery of theorems as well as to make him/her think about the inner beauty of classical solutions and about some less known results.

A correspondence between varieties and ideals enables us to pass between geometry and algebra. This gives us a big power. Given a problem, we can decide whether to solve it classically using the geometric properties of a variety or by means of the tools of computer algebra with the use of the theory of automated theorem proving.

The book is suitable for those who are interested in geometry or algebra, in classical solutions or automated theorem proving and elimination of variables. It is accessible to undergraduate and graduate students yet of interest to experts.

I am indebted to many people who helped me in the course of work on the book. I am especially grateful to Tomas Recio (Univ. of Cantabria in Santander) for encouragement and help. I also wish to thank Dongming

Wang (Univ. Pierre et Marie Curie in Paris) for his support. Jaroslav Hora and Miroslav Lávička (Univ. of West Bohemia in Pilsen) and Naďa Stehlíková (Charles University in Prague) deserve thanks for careful reading of the manuscript and valuable suggestions and recommendations which improved the text. It is my pleasure to thank Ms. Ji Zhang and Mei Kee Hoe, production editors from World Scientific, who helped me during the preparation of the book.

I would like to devote special thanks to my wife Ilona for her patience with me during the work on the book.

P. Pech

Contents

Chapter 1

Introduction

In the last third of the 20th century efficient methods in automatic theorem proving in geometry were developed. By means of this theory hundreds of non-trivial theorems have been proved and even discovered. A correspondence between varieties and ideals gives us the ability to solve a given problem classically by means of the properties of geometric objects or by computer algebra methods based on the theory of automated theorem proving.

In the seminar which I led in the last years at the University of South Bohemia we solved on computers a number of problems of elementary geometry by the theory of automatic theorem proving. Students who attended this optional seminar were mostly at their 4th year of study, i.e., they had basic knowledge of geometry. Both by computers and in a classical way we investigated a number of tasks — the formula of Heron for the area of a triangle and its generalization — the formula of Brahmagupta for the area of an inscribed quadrilateral in terms of its lengths of sides, the formula of Staudt, Simson–Wallace theorem and its generalization, Napoleon's theorem and further similar problems.

A classical (or synthetic) method gives a better insight into the given geometrical situation which also enables a better understanding of a problem and shows the beauty of geometry. On the other side by computer algebra we can solve complex problems which are difficult to solve by classical approach. Using computer algebra we can carry out automatic proving theorems, automatic derivation and automatic discovery of new theorems. Whereas under automatic derivation we understand finding a statement which follows from the given assumptions, automatic discovery means searching for additional conditions which are necessary to add to the given assumptions so that the statement, which is not in general true, becomes valid. Automatic derivation is sometimes considered as a special case of automatic discovery. By computers we are also able to make constructions which are difficult to construct by the ruler and compass, etc.

First we give a brief overview of the theory of automatic theorem proving. To those who are interested in deeper study I recommend the book D. Cox, J. Little, D. O'Shea: Ideals, Varieties, and Algorithms [24] and D. Wang: Elimination Practice. Software Tools and Applications [136]. See also X. S. Gao, D. Wang (eds.): Mathematics Mechanization and Applications [43], where the overview of mechanizing mathematical activities, such as calculation, reasoning and discovery is given.

This book can be useful for those who are interested in geometry and proving theorems. In a few chapters which follow after the exposition of necessary theory on automatic theorem proving, both well-known and less known problems are solved — I called them stories — with the author's effort to describe their solution from the beginning to present time.

The first story "Generalization of the formula of Heron" in Chapter 3 deals with a generalization of Heron's and Brahmagupta's formulas for the area of a triangle and an inscribed quadrilateral in terms of the lengths of sides on an inscribed pentagon. This problem, which was solved in 1994 by D. P. Robbins [112], is investigated in this book by elimination of variables.

In the second story "Simson–Wallace Theorem" (Chapter 4) this well-known theorem of planar geometry is generalized into three dimensional space. First known generalizations of the Simson–Wallace theorem in a plane, from which particularly the Guzmán's generalization deserves attention, are demonstrated. Furthermore two spatial analogies of this theorem, which lead to the cubic surfaces with interesting properties, are shown.

Various generalizations of Ceva's, Menelaus' and Euler's theorems in a plane and space are studied in the part "Transversals in a polygon" (Chapter 5). In it, the power of the computer approach which searches for new formulas, and the power of the traditional yet possibly neglected area method (area principle), are shown.

The Napoleon's theorem and its generalization — Petr's theorem — and their planar and spatial analogies are given in the story "Petr–Douglas–Neumann theorem" (Chapter 6). As the name says, this theorem is connected with the well-known Czech mathematician K. Petr, a professor of the Charles University in Prague, who first published the theorem in 1905. Several special cases of the PDN theorem are solved — Thébault's theorem, Finsler–Hadwiger theorem, theorem of Finney, Van Aubel's theorem etc. Kiepert hyperbola, which is closely connected with Napoleon's theorem, is investigated as a locus of points of a given property.

In "Geometric inequalities" (Chapter 7), the inequality between lengths of sides and diagonals of a polygon in a plane and space is gradually generalized. The base is the well-known parallelogram law which played an important role in the thirties of the last century when it was shown that Banach space, in which parallelogram law holds, is the Hilbert space. Further the inequality between the sum of squares of sides of a polygon and its area is investigated. Then the Neuberg–Pedoe inequality for two triangles and the discrete case of Wirtinger's inequality are explored. The isoperimetric inequality for polygons concludes this family of inequalities. All these inequalities are proved using the elimination of variables and the sum of squares decomposition of polynomials.

Another method of proving inequalities using computer is shown on the well-known Euler's inequality between the radii of the circumcircle and the incircle of a triangle.

The next story "Regular polygons" (Chapter 8) is devoted to problems connected with regular polygons. Although it seems that all important things about regular polygons have been said, this is not the case. In 1969 two chemists visited the well-known mathematician B. L. Van der Waerden and stated that according to their investigations an equilateral and equiangular pentagon in a space must necessarily be planar. During a short period Van der Waerden and a number of further mathematicians proved that a regular polygon with an odd number of vertices always has even dimension. It is proved that regular pentagons and heptagons span spaces of even dimension.

Exploring properties of regular polygons in E^d, we might find ourselves in a situation where it is necessary to show that the conclusion polynomial belongs to the radical $\sqrt{I}$ of an ideal I. It is shown that in this case the respective ideal I is a proper subset of the radical $\sqrt{I}$ and we have to apply the stronger criterion, i.e. to show that c belongs to the radical $\sqrt{I}$.

In "Miscellaneous" (Chapter 9) a few problems of various parts of geometry which seem to be of interest are investigated. Especially the first two parts — "Non-elementary constructions" and "Loci of points of given properties" — are important both from the geometry and algebra point of view. Whereas loci of points belong to frequent issues of the theory of elimination, see [136], non-elementary constructions are not often mentioned in computer algebra topics. By means of a computer we are able to solve even such problems which are (Euclidean) unsolvable by a ruler and compass and were taboo in the past. Further the theorem of Viviani and the theorem of Gauss which is also known as the theorem on a complete quadrilateral are investigated. The theorem of Viviani is generalized using the method of automatic discovery based on the extension of the conclusion variety. Proving the theorem of Gauss automatically and classically we would realize that sometimes one method seems to be simpler than the other one.

At the end of this chapter is a list of the most common problems that students might encounter in the course of the seminar on automatic theorem proving at the university.

Classical solutions to all problems are mostly given. If a classical solution is missing it is likely caused by the fact that the author does not know it.

Computations were done in computer algebra system CoCoA[1]. Sometimes the programs Singular and Maple were used. Given computations can be carried out with another common mathematical software using elimination of variables based on Gröbner bases computation as well (Mathematica, Derive, ...).[2]

Figures are drawn in dynamic geometry software Cabri II Plus. All computations were done on a computer Intel Pentium 2.00GHz/1536MB RAM.

[1]Software CoCoA is freely distributed at the address `http://cocoa.dima.unige.it`

[2]The book [136] contains the application module GEOTHER working under Maple, which provides an environment for handling and proving theorems in geometry automatically.

Chapter 2

Automatic theorem proving

2.1 History

The theory of automatic theorem proving involves automatic proving, automatic derivation and automatic discovery of mathematical theorems, i.e., proving, deriving and discovering by computer. Thus it is not only about proving theorems as the name says. However, this name is internationally approved for the theory which was first used to prove theorems. If automatic deriving or discovering is considered we will stress this fact.

When a basic book Mechanical Geometry Theorem Proving written by Chinese mathematician S. Ch. Chou appeared in 1987 [22], American mathematician Larry Woss wrote in its preface:

"When computers were first conceived, then designed, and finally implemented, few people (if any) would have conjectured that in 1987 computer programs would exist capable of proving theorems from diverse areas of mathematics. Even further, if a person at the inception of the computer age had seriously predicted that computer programs would be used to occasionally answer open questions taken from mathematics, that person would have received at best a polite smile."

We think that these words well describe the opinions of many people on computers even today, twenty years later.

Results from mathematical logic and A. Tarski's works [130] from the thirties of the 20th century predetermined further development in the theory of automatic theorem proving. A basic work by Chinese mathematician W. Wu [143] from 1978 is based on procedures of pseudodivision and triangulation of a system of polynomial equations. The ideas of Wu were followed by another Chinese mathematician S. Ch. Chou. His book Mechanical Geometry Theorem Proving [22], in which 500 examples from elementary geometry are solved, became the best known.

In the seventies another method which is based on Gröbner bases of an ideal [18] appears. This method was further developed by D. Kapur [58] and others, who compared both basic methods — the method by Wu and the method based on Gröbner bases computation.

In this book we will use the elimination method which is based on the theory of Gröbner bases of ideals (see [24, 136, 135]).

The community that works on the subject of automatic theorem proving is still discussing some fundamental notions, such as the idea of truth and the fact that different approaches can lead to considering that a theorem is true or that is false. Refer to [23, 19, 104] and some of their references for more details.

2.2 Basic notions from algebra

This chapter contains those notions from algebra which are the key stones in the theory of automatic proving theorems. Even though the choice of notions from algebra is not complete, it is sufficient for our purposes. It serves as a guidance for orientation in the given theory. For a deeper study [24, 22] are recommended.

Let K denote a commutative field and $K[x_1, \ldots, x_n]$ a ring of polynomials of n indeterminates with coefficients in K. For K we will mostly take the fields of real numbers $\mathbb{R}$ or complex numbers $\mathbb{C}$.

Definition 2.1. Let $h_1, \ldots, h_r$ be polynomials in $K[x_1, \ldots, x_n]$. Then the set

$$V(h_1, \ldots, h_r) = \{(a_1, \ldots, a_n) \in K^n; h_i(a_1, \ldots, a_n) = 0 \text{ for } 1 \leq i \leq r\} \tag{2.1}$$

is called the algebraic (or afinne) variety defined by $h_1, \ldots, h_r$.

We also say that an algebraic variety $V(h_1, h_2, \ldots, h_r)$ is a zero set of polynomials $h_1, \ldots, h_r$. Let $K = \mathbb{R}$ and consider points $[x, y]$ in the plane $\mathbb{R}^2$. Then $V(x^2 - y)$ is the parabola, $V(x^2 - y, x - y)$ is the intersection of the parabola $x^2 - y = 0$ and the straight line $x - y = 0$, i.e., points $[0, 0]$, $[1, 1]$, etc. An example of a variety in the space $\mathbb{R}^3$ is the sphere $V(x^2 + y^2 + z^2 - 1)$ with the radius 1 centered at the origin.

The next important notion is an ideal.

Definition 2.2. A subset $I \subset K[x_1, \ldots, x_n]$ is an ideal if it satisfies:

1) $0 \in I$.
2) *If* $f, g \in I$, *then* $f + g \in I$.
3) *If* $f \in I$ *and* $c \in K[x_1, \ldots, x_n]$, *then* $cf \in I$.

Hence an ideal is a set of polynomials which is closed under addition and multiplication by an arbitrary polynomial from the ring of polynomials $K[x_1, \ldots, x_n]$. We have a theorem:

Theorem 2.1. *Let* $h_1, \ldots, h_r$ *be polynomials in* $K[x_1, \ldots, x_n]$. *Then the set*

$$(h_1, \ldots, h_r) = \{g_1 h_1 + \cdots + g_r h_r;\ g_1, \ldots, g_r \in K[x_1, \ldots, x_n]\} \tag{2.2}$$

is an ideal. We call $(h_1, \ldots, h_r)$ *the ideal generated by* $h_1, \ldots, h_r$.

We say that an ideal I is *finitely generated* if there exist $h_1, \ldots, h_r \in K[x_1, \ldots, x_n]$ such that $I = (h_1, \ldots, h_r)$. We say that $h_1, \ldots, h_r$ form a *basis* of I. An ideal I may have many different bases. One type of a basis with special properties, called *Gröbner basis,* is very useful and we will apply it in our considerations.

The ideal $(h_1, \ldots, h_r)$ contains all linear combinations $g_1 h_1 + \cdots + g_r h_r$ of polynomials $h_1, \ldots, h_r$, where coefficients are polynomials $g_1, \ldots, g_r$ from $K[x_1, \ldots, x_n]$. The situation is similar to the definition of a vector subspace generated by vectors $h_1, \ldots, h_r$. Then an arbitrary vector of a subspace which is generated by vectors $h_1, \ldots, h_r$ may be expressed as a linear combination $g_1 h_1 + \cdots + g_r h_r$, where $g_1, \ldots, g_r$ are scalars from a field K.

One of the major issues of algebraic geometry is the so-called "Ideal Membership Problem," i.e., the problem of how to recognize that a polynomial h belongs to the ideal generated by polynomials $h_1, \ldots, h_r$. This problem was solved in the last third of the 20th century by means of Gröbner bases of an ideal (see [18, 24]).

As we said a Gröbner basis of an ideal I is a basis with specific properties. One of the main properties of a Gröbner basis is described in the following theorem:

Theorem 2.2 (Ideal Membership). *Let* $G = \{g_1, \ldots, g_s\}$ *be a Gröbner basis of an ideal* $I \subset K[x_1, \ldots, x_n]$ *with a fixed monomial ordering and let* $h \in K[x_1, \ldots, x_n]$. *Then* $h \in I$ *if and only if the remainder* r *on division of* h *by polynomials from* G *is zero.*

Taking a basis G of an ideal I which is not a Gröbner basis, we get on division of $h \in K[x_1, \ldots, x_n]$ by the elements of G different remainders dependent on the order of divisors. A Gröbner basis has such a property that on division of h by the elements of G no matter how the elements of G are ordered, the remainder is always the same.

Computation of a Gröbner basis of an ideal may be carried out by the so-called *Buchberger's algorithm* which is due to B. Buchberger [18]. This algorithm is implemented in most of the mathematical programs. For example in the program CoCoA we enter the command `GBasis(I)`.

When investigating whether a polynomial h lies in an ideal I or not, it suffices to compute a Gröbner basis G of I with some prescribed ordering and then to find the remainder r on division of h by the elements of G. If the remainder $r = 0$ then $h \in I$, if $r \neq 0$ then $h \notin I$.

The remainder r is often called the *normal form of h*. The algorithm for computing the normal form of h with respect to an ideal I is implemented in most mathematical programs as well. For instance in CoCoA this command has the form `NF(h,I)`.

When solving a system of algebraic equations, we successively eliminate variables in a similar manner as by using the Gaussian elimination procedure to solve linear systems of equations. When eliminating variables in an ideal I we get, in accordance with the following definition, the so-called *elimination ideal*:

Definition 2.3. Given an ideal $I = (h_1, \ldots, h_s) \subset K[x_1, \ldots, x_n]$. Then the r-th elimination ideal I_r is the ideal

$$I_r = I \cap K[x_{r+1}, \ldots, x_n].$$

For a Gröbner basis of the elimination ideal I_r we get the theorem:

Theorem 2.3. *Let $I \subset K[x_1, \ldots, x_n]$ be an ideal and let G be a Gröbner basis of I with respect to the lexicographic ordering $x_1 > x_2 > \cdots > x_n$. Then for every $0 \leq r \leq n$ the set*

$$G_r = G \cap K[x_{r+1}, \ldots, x_n]$$

is a Gröbner basis of the r-th elimination ideal I_r.

Computing a Gröbner basis of the elimination ideal is accessible by the command `Elim` or by its analogies which are involved in most mathematical programs. This command is usually applied to solving systems of algebraic equations.

Theorem 2.4. *Let $V \subset K^n$ be an algebraic variety. Then the set*

$$I(V) = \{h \in K[x_1, \ldots, x_n]; h(a_1, \ldots, a_n) = 0 \text{ for all } (a_1, \ldots, a_n) \in V\} \tag{2.3}$$

is an ideal. We call $I(V)$ the ideal of V.

It is easy to verify that the ideal $I(V)$ of V is an ideal. Let us introduce another notion which we will need.

Definition 2.4. Let $I \subset K[x_1, \ldots, x_n]$ be an ideal. The set

$$V(I) = \{(a_1, \ldots, a_n) \in K^n; h(a_1, \ldots, a_n) = 0 \text{ for all } h \in I\} \tag{2.4}$$

is called the variety of I.

Hence, if $I = (h_1, \ldots, h_r)$ then $V(I) = V(h_1, \ldots, h_r)$. The last definition is a consequence of the Hilbert basis theorem which says that every ideal $I \subset K[x_1, \ldots, x_n]$ has a finite set of generators.

Given an ideal I we can assign to it its zero set — the variety $V(I)$. To this zero set $V(I)$ we assign its ideal $I(V(I))$. The question now arises: what does this ideal look like?

The answer to this question is given by the famous Hilbert's *Nullstellensatz.* Before we say this theorem we will give a definition of a radical of an ideal.

Definition 2.5. Let $I \subset K[x_1, \ldots, x_n]$ be an ideal. Then the set

$$\sqrt{I} = \{f;\ f^k \in I \text{ for some integer } k \geq 1\}$$

is what we call the radical of an ideal I.

Obviously $I \subset \sqrt{I}$ for $f \in I$ implies $f^1 \in I$ and from here $f \in \sqrt{I}$. Hence the radical $\sqrt{I}$ contains the ideal I. It is the interesting fact that the radical $\sqrt{I}$ of I is again an ideal:

Theorem 2.5. *Let I be an ideal in $K[x_1, \ldots, x_n]$. Then the radical $\sqrt{I}$ of I is the ideal in $K[x_1, \ldots, x_n]$.*

The Hilbert's theorem is as follows:

Theorem 2.6 (Nullstellensatz). *Let I be an ideal in $K[x_1, \ldots, x_n]$ where K is an algebraically closed field. Then*

$$I(V(I)) = \sqrt{I}. \tag{2.5}$$

The following theorem says how we can find out whether a polynomial h lies in the radical $\sqrt{I}$ of an ideal I or not.

Theorem 2.7 (Radical Membership). *Let K be an arbitrary field and let $I = (h_1, \dots, h_r) \subset K[x_1, \dots, x_n]$. Then $h \in \sqrt{I}$ if and only if the constant polynomial* 1 *belongs to the ideal $J = (h_1, \dots, h_r,\ ht - 1) \subset K[x_1, \dots, x_n, t]$.*

By last definitions and theorems we can assign to every ideal I the variety $V(I)$ of an ideal I

$$I \rightarrow V(I)$$

and conversely, to every variety V we can assign the ideal $I(V)$ of a variety V

$$V \rightarrow I(V).$$

Now a theorem which gives a correspondence between algebraic varieties and ideals follows [24]:

Theorem 2.8 (The Ideal-Variety Correspondence). *Let k be an arbitrary field.*
a) The maps

$$I: \text{ algebraic varieties } \rightarrow \text{ ideals}$$

and

$$V: \text{ ideals } \rightarrow \text{ algebraic varieties}$$

are inclusion-reversing, i.e., if $I_1 \subset I_2$ are ideals, then $V(I_1) \supset V(I_2)$. Similarly, if $V_1 \subset V_2$ are varieties, then $I(V_1) \supset I(V_2)$. Furthermore for any variety V the equality $V(I(V)) = V$ holds, so that I is always one-to-one.
b) If k is algebraically closed, and if we consider radical ideals, then the maps

$$I: \text{ algebraic varieties } \rightarrow \text{ radical ideals}$$

and

$$V: \text{ radical ideals } \rightarrow \text{ algebraic varieties}$$

are inclusion-reversing bijections which are inverses of each other.

The previous theorem needs to be clarified. Whereas the map I is always one-to-one, i.e. for instance, if $V_1 \neq V_2$, then $I(V_1) \neq I(V_2)$, the map V is not one-to-one — different ideals can give the same variety. For example, the different ideals $I_1 = (x, y)$ and $I_2 = (x^2, y^3)$ in $k[x, y]$ have the same variety $V(I_1) = V(I_2) = \{[0,0]\}$. This drawback can be removed if we consider radicals of ideals instead of ideals. In our case obviously $\sqrt{I_1} = \sqrt{I_2}$.

In addition, if the field k is not algebraically closed, then the following situation may happen: for different ideals $J_1 = (x^2+1)$, $J_2(x^2+2)$ in $\mathbb{R}[x]$ the corresponding varieties $V(J_1), V(J_2)$ are empty. We remove this second drawback if we are working over an algebraically closed field k.

Theorem 2.8 enables us to pass between algebra and geometry which gives us big power. Any problem about varieties (geometry) can be solved as an algebraic problem about radical ideals (algebra) and conversely. We can understand this as the ability to switch between classical and computer algebra methods of solving problems in geometry (and algebra) which in fact is the name of this book.

2.3 Automatic theorem proving

Let K be a field of characteristic 0, for instance the field of rational numbers $\mathbb{Q}$, and L an algebraically closed field containing K, for instance the field of complex numbers $\mathbb{C}$. Further denote by $K[x_1, \ldots, x_n]$ the ring of polynomials of n indeterminates $x = (x_1, \ldots, x_n)$ with coefficients in the field K.

Automatic theorem proving deals with geometric statements which are of the form $H \Rightarrow T$, where H is a set of hypotheses and T is a set of conclusions (theses).

The basic steps of automatic theorem proving are as follows:

(1) Introduction of a coordinate system.
(2) Algebraic formulation of a problem.
(3) Proof of a statement.
(4) Searching for additional conditions if necessary.

Now we will describe individual steps in detail.

According to the nature of a problem we *introduce an appropriate coordinate system* in the plane, in 3D space or in the space of higher dimension (Cartesian, affine, projective, ...). The choice of a suitable coordinate system is very important. We always try to choose such a coordinate system so that the respective algebraic equations are as simple as possible. Different choices of a coordinate system lead to a different complexity of solving the problem.

After introduction of a coordinate system we carry out *algebraic formulation* of a problem. We translate geometric properties of objects into algebraic relations of *equality type,* in which algebraic formulation involves only polynomial equations (=) and inequations ($\neq$). This phase is characterized by establishing a set of hypotheses H which have the form of algebraic equations, i.e., they are expressed in the form of polynomials

$$h_1(x_1, \ldots, x_n) = 0,\ h_2(x_1, \ldots, x_n) = 0, \ldots, h_r(x_1, \ldots, x_n) = 0$$

and a set of conclusions T

$$c_1(x_1, \ldots, x_n) = 0, c_2(x_1, \ldots, x_n) = 0, \ldots, c_s(x_1, \ldots, x_n) = 0,$$

where $h_1, \ldots, h_r, c_1, \ldots, c_s \in K[x_1, \ldots, x_n]$. Without loss of generality assume that the set of conclusions T contains only one conclusion[1] which we denote by c, i.e.

$$c(x) = 0.$$

Hence the result of the second step is the algebraic form of a statement

$$\forall x \in L^n, \quad h_1(x) = 0, \ldots, h_r(x) = 0 \quad \Rightarrow \quad c(x) = 0. \tag{2.6}$$

Algebraic formulation of a problem has a tremendous influence on the result of investigation. Working in the field of complex numbers we cannot use inequalities which is a consequence of the well-known fact that the field of complex numbers cannot be ordered. Thus all geometric properties of objects must be expressed by equations or inequations. This often makes it difficult to solve problems which depend on orientation, for instance inner or outer points of objects, etc. This leads to non-uniqueness from which many problems arise.

Further we should keep algebraic equations and inequations in a simple form, e.g., the degree of polynomials and number of variables should be as low as possible. For instance we can decide whether to use the ratio of points or the classical distance to express the position of an arbitrary point of a segment AB with respect to the endpoints A, B.

We should pay attention to the exact description of geometric properties by their translation into algebraic equations [68] as well. If the translation is not exact then problems with the determination of additional conditions arise in proofs, etc.

The third step in automatic theorem proving is the *proof* of the statement (2.6) (Fig. 2.1).

[1] If the set of conclusions T consists of more equations then we investigate every equation separately.

Definition 2.6. Let us consider an ideal $I = (h_1, \ldots, h_r)$. The statement (2.6) is generally true if the hypotheses variety $V(H)$ is a subset of the conclusion variety $V(T)$, that is,

$$V(H) \subset V(T). \tag{2.7}$$

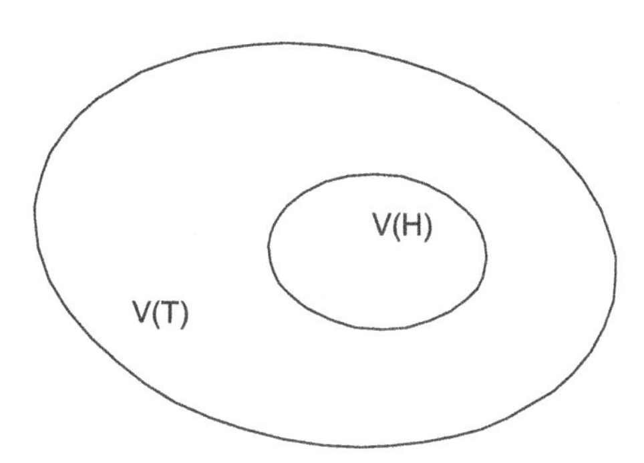

Fig. 2.1 $H \Rightarrow T \Leftrightarrow V(H) \subset V(T)$

Note that the hypotheses variety $V(H)$ is the set of all solutions of the system of equations $h_1 = 0, \ldots, h_r = 0$, i.e.,

$$V(H) = \{x \in L^n;\ h_1(x) = 0, \ldots, h_r(x) = 0\}. \tag{2.8}$$

Hence $V(H)$ is a zero set of polynomials $h_1 = 0, \ldots, h_r = 0$. In accordance with the theory above we can write $V(I)$, where $I = (h_1, \ldots, h_r)$. The conclusion variety $V(T)$ is a set of all solutions of the equation $c = 0$, i.e.,

$$V(T) = \{x;\ c(x) = 0\}. \tag{2.9}$$

The following theorem gives an outline on how to show the validity of (2.7). The theorem reads [101]:

Theorem 2.9. *The following statements are equivalent:*

a) The statement (2.6) is generally true.
b) $c \in \sqrt{(h_1, \ldots, h_r)}$.
c) $1 \in (h_1, \ldots, h_r, ct - 1) \subset K[x_1, \ldots, x_n]$.

Note that $V(H) \subset V(T)$ is equivalent to $V(H) \setminus V(T) = \emptyset$ (Fig. 2.2). This means that $V(H, ct - 1) = \emptyset$ or the Gröbner basis of the ideal $(h_1, h_2, \ldots, h_r, ct - 1)$ is $\{1\}$ using any term order. This proves the part c) of Theorem 2.9 above.

By the previous theorem we determine the normal form of the constant polynomial 1 with respect to the Gröbner basis of the ideal $J = (h_1, \ldots, h_r, ct - 1)$ for some prescribed order of variables. In the computer algebra system CoCoA we use the command `NF(1,J)`. If the answer is 0 then it means that $1 \in J$ and (2.6) is generally true.

Another way of determining whether the constant 1 belongs to the ideal J is to compute the Gröbner basis of the ideal J using any term order. This is especially useful in the case if we use the software which does not automatically compute the normal form using the Gröbner basis of a set of generators.

If the answer is not zero then the statement (2.6) is not generally true and it is necessary to look for additional conditions (see subsequent pages).

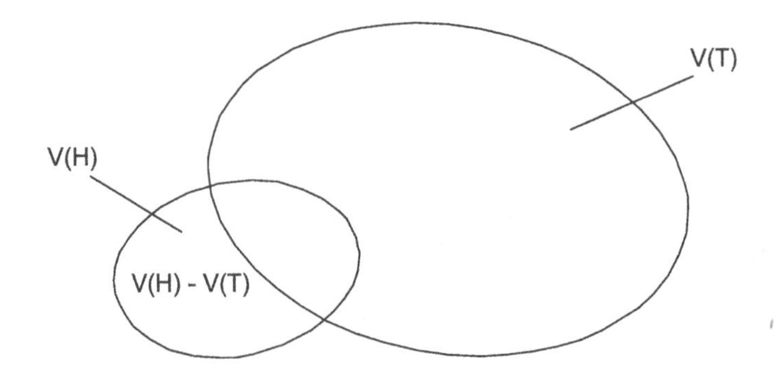

Fig. 2.2 $H \Rightarrow T \Leftrightarrow V(H) \subset V(T) \Leftrightarrow V(H) \setminus V(T) = \emptyset \Leftrightarrow V(H, ct - 1) = \emptyset$

Remark 2.1. In practice it is usually sufficient to show that a conclusion polynomial c is in the ideal $I = (h_1, \ldots, h_r)$ since in most cases $I = \sqrt{I}$. One computes the normal form of a polynomial c with respect to the Gröbner basis of the ideal I with some prescribed order of variables using the command `NF(c,I)`. If we get the answer 0 then a statement is generally true. If we do not get zero then it is necessary to apply the previous stronger criterion and find out whether 1 is an element of the ideal $J = I \cup \{ct - 1\}$. In Chapter 8 we will encounter the case when the set of hypotheses is not described by the radical ideal, i.e. $I \subsetneqq \sqrt{I}$ and c is not in the ideal I, but it is an element of the radical $\sqrt{I}$.

The easiest way to demonstrate automatic theorem proving is by use of an example. We will solve the following problem.

Example 2.1. Prove that altitudes of a triangle are concurrent.

First, we will choose an appropriate coordinate system so that the relations by means of which a geometric situation will be described are as simple as possible (Fig. 2.3). As this is a Euclidean problem (we use orthogonality) we introduce a Cartesian coordinate system. Denote by $A = [0, 0]$, $B = [a, 0]$,

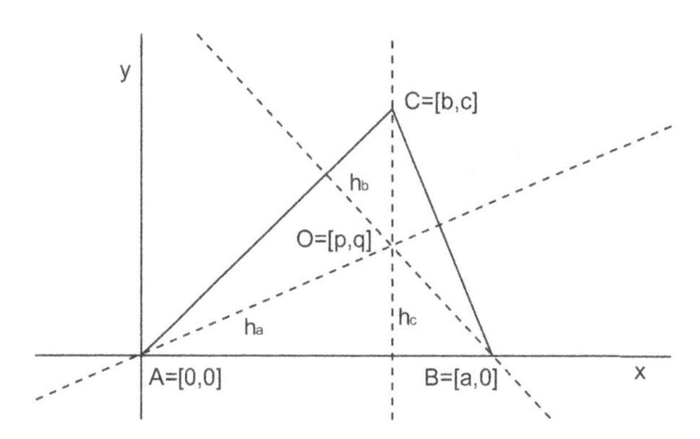

Fig. 2.3 Altitudes of a triangle ABC are concurrent — computer proof

$C = [b, c]$ the vertices of a triangle ABC. Now we express equations of the altitudes h_a, h_b, h_c of ABC:

$h_a : (b - a)x + cy = 0,$
$h_b : bx + cy - ab = 0,$
$h_c : x - b = 0.$

Suppose that the altitudes h_b and h_c are concurrent at a point $O = [p, q]$, then it holds:

$O \in h_b \Leftrightarrow h_1 : bp + cq - ab = 0,$
$O \in h_c \Leftrightarrow h_2 : p - b = 0.$

We want to show that the conclusion z — the altitude h_a contains the point O — is consequently fulfilled:

$O \in h_a \Leftrightarrow z : (b - a)p + cq = 0.$

Hence we are to prove the following statement:

$$\forall p, q, \quad bp + cq - ab = 0,\ p - b = 0 \quad \Rightarrow \quad (b - a)p + cq = 0. \qquad (2.10)$$

In this very simple case we are able to show that the statement (2.10) is true even by hand — without using computer. Indeed, realize the equality

$$(b-a)p + cq = 1 \cdot (bp + cq - ab) - a \cdot (p - b). \tag{2.11}$$

We expressed the conclusion polynomial $(b-a)p+cq$ as a linear combination of hypotheses polynomials $bp + cq - ab$ and $p - b$. From the validity of the equations $bp + cq - ab = 0,\ p - b = 0$ and from (2.11) the validity of the equation $(b - a)p + cq = 0$ follows.

We showed that the polynomial $(b - a)p + cq$ belongs to the ideal $I = (bp + cq - ab, p - b)$. In the program CoCoA we enter

```
Use R::=Q[abcpq];
I:=Ideal(bp+cq-ab,p-b);
NF((b-a)p+cq,I);
```

and get the answer 0 which means that `NF((b-a)p+cq,I)=0` and the statement (2.10) is generally true. A "computer" proof is finished.

Usually the situation is not as simple as in the presented example. In most cases it is necessary to add to given assumptions so-called non-degeneracy conditions.

In the fourth step of automatic theorem proving we concentrate on *searching for additional conditions.*

If we do not manage to prove that the statement (2.6) is generally true, it still does not mean that this statement is not valid. It can happen that a statement is not generally true because of missing *non-degeneracy conditions* which we include among so-called *additional* or *subsidiary* or *side* conditions. This step is characterized by searching for such conditions under which a statement is meaningless, for instance, three vertices of a triangle are collinear, a segment of zero length, circles of zero radii, etc. These situations must be ruled out. Non-degeneracy conditions have a form of inequations

$$g_1(x_1, \ldots, x_n) \neq 0, \ldots,\ g_s(x_1, \ldots, x_n) \neq 0 \tag{2.12}$$

and are expressed only by independent variables, i.e., by those variables which do not depend on other variables. We will denote them $x_1, \ldots, x_m$ and call them *independent* variables. Remaining variables $x_{m+1}, \ldots, x_n$ are *dependent* for they depend on independent variables. By means of independent variables we usually determine basic points of an object.

Non-degeneracy conditions (2.12) can be found in the following way. In the ideal $J = I \cup \{ct - 1\}$ we eliminate a slack variable t and all *dependent* variables. In this newly formed elimination ideal degeneracy conditions $g_1 = 0, \ldots, g_s = 0$ may occur (if they exist). Elimination of variables in CoCoA can be executed by the command `Elim`. If we involve conditions (2.12) into the statement (2.6), we get

$$\forall x \in L^n, h_1(x) = 0, \ldots, h_r(x) = 0, g_1(x) \neq 0, \ldots, g_s(x) \neq 0 \Rightarrow c(x) = 0. \tag{2.13}$$

In order to prove a statement in the form (2.13) it suffices to show that the constant polynomial 1 belongs to the ideal

$$I' = (h_1, \ldots, h_r, g_1 t_1 - 1, \ldots, g_s t_s - 1, ct - 1),$$

where $t_1, \ldots, t_s, t$ are slack variables. Note here an algebraic translation of, for instance, the inequation $g_1 \neq 0$, is $g_1 t_1 - 1 = 0$. Indeed, if $g_1 = 0$ then we get $-1 = 0$ and zero set of an ideal I' is empty.[2]

Definition 2.7. If the normal form of the constant polynomial 1 with respect to an ideal I' equals zero then we say that the statement (2.6) is generically true.

Most geometric statements are generically true. Non-degeneracy conditions of investigated objects are usually not involved in the statements. We tacitly suppose that a triangle is a real triangle and not a segment, that a segment is a real segment and not a pair of identical points, that a circle has the non-zero radius, etc. We say that such objects are *generic*. In the instances that objects are not generic a theorem does not often hold or is meaningless.

Let us show the use of non-degeneracy conditions on the example which we have already dealt with.

Prove that altitudes of a triangle are concurrent.

Unlike the previous procedure we assume that now the altitudes h_a and h_b intersect at the point $O = [p, q]$ (instead of the altitudes h_b and h_c). We want to prove that the remaining altitude h_c also passes through the point O (Fig. 2.3). This is the same as to show that from the validity of equations $(b - a)p + cq = 0$ and $bp + cq - ab = 0$ the conclusion $p - b = 0$ follows. Now our statement has the form

$$\forall p, q, \quad (b - a)p + cq = 0, \; bp + cq - ab = 0 \quad \Rightarrow \quad p - b = 0. \tag{2.14}$$

We explore the normal form of the conclusion polynomial $p - b$ in the ideal $I = ((b - a)p + cq, bp + cq - ab)$. In CoCoA we enter

[2]There are also other options (such as computing the saturation) which yield different results. See the discussion and counterexamples in [27].

```
Use R::=Q[abcpq];
I:=Ideal((b-a)p+cq,bp+cq-ab);
NF(p-b,I);
```

The answer $p - b$ means that the polynomial $p - b$ does not belong to the ideal I. Therefore we will use a stronger criterion to investigate if the polynomial $p-b$ is an element of the radical $\sqrt{I}$. By Theorem 2.7 we enter

```
Use R::=Q[abcpqt];
J:=Ideal((b-a)p+cq,bp+cq-ab,(p-b)t-1);
NF(1,J);
```

and get the answer 1 which means that the statement (2.14) is not generally true.

Now we determine non-degeneracy conditions (if there are any). To find these conditions we add a negated conclusion polynomial $(p-b)t-1$ to the ideal $I = ((b-a)p+cq, bp+cq-ab)$ and obtain the ideal $J = ((b-a)p+cq, bp+cq-ab, (p-b)t-1)$, where t is a slack variable. By adding the polynomial $(p-b)t-1$ to the ideal I, we assume that the conclusion does not hold, whereas all other assumptions are preserved (as by the proof by contradiction). In fact we are asking: "What is the reason of invalidity of a conclusion?" We eliminate *dependent* variables p, q, t in the ideal J. The resulting elimination ideal contains only polynomials in remaining *independent* variables a, b, c. We enter

```
Use R::=Q[abcpqt];
J:=Ideal((b-a)p+cq,bp+cq-ab,(p-b)t-1);
Elim(p..t,J);
```

and get the condition $a = 0$. Geometrically it means that the vertices A and B of a triangle ABC coincide. We will exclude this case by adding a non-degeneracy condition $at - 1 = 0$ to the ideal I. In this way we get the ideal $K = I \cup \{at - 1\}$.

The whole process will now repeat. The only change is that instead of the ideal I we explore the ideal K.

```
Use R::=Q[abcpqt];
K:=Ideal((b-a)p+cq,bp+cq-ab,at-1);
NF(p-b,K);
```

The answer 0 means that the statement "Altitudes of a triangle are concurrent" is generically true.

Why it is necessary to exclude the case $a = 0$ can be seen from the equality

$$p - b = -\frac{1}{a}((b-a)p + cq) + \frac{1}{a}(bp + cq - ab),$$

where the polynomial $p - b$ is expressed as a linear combination of polynomials $(b-a)p + cq$ and $bp + cq - ab$ with one exception, namely $a = 0$.

Remark 2.2.
1) We saw two almost identical solutions of the same problem. In the first solution we did not need to investigate non-degeneracy conditions, but this was not the case in the second solution. Therefore it is important to realize the role of strategy and a choice of a coordinate system, see point 2) of this remark.

2) Arisen problems can be removed by the following choice of the coordinates of vertices of a triangle $A = [0,0]$, $B = [1,0]$, $C = [b,c]$, where we put $a = 1$ (the unit of the length can be chosen). By this choice of the coordinate system both strategies above lead to the same result — the general validity of this statement.

As in this whole text we use both computer and classical proofs when solving problems, now a classical proof of the fact that the altitudes of a triangle are concurrent follows.

Through the vertices A, B, C of a triangle ABC we lead straight lines which are parallel to opposite sides BC, AC, AB respectively. We get a new triangle $A'B'C'$ in which the altitudes h_a, h_b, h_c of a triangle ABC are side bisectors (Fig. 2.4). Now it suffices to show that side bisectors of the triangle $A'B'C'$ are concurrent. This is easy since from $|A'O| = |B'O|$ and $|B'O| = |C'O|$ we get $|A'O| = |C'O|$.

To prove the statement we can also use the theorem of Ceva, see Chapter 5 "Transversals in a polygon".

A classical proof that we just showed had, besides many positive qualities, one major drawback — we needed *a key idea* which leads to the solution of a problem. Sometimes it can happen that no key idea will come.

2.4 Automatic derivation

We will deal with *automatic derivation* of statements which we mostly distinguish from automatic discovery of theorems (see the next section). By

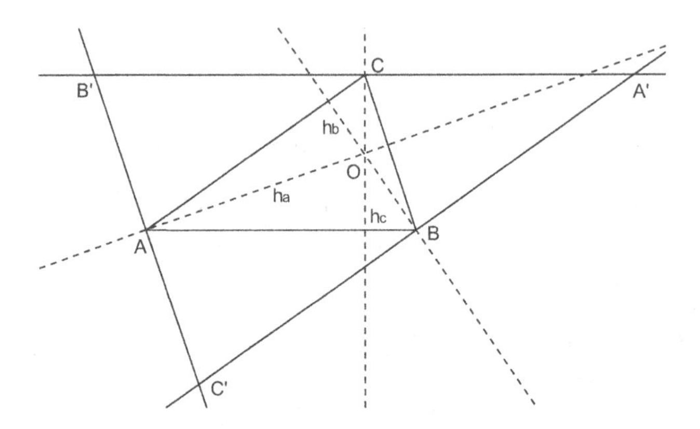

Fig. 2.4 Altitudes of a triangle are concurrent — classical proof

automatic derivation of theorems [102] we mean finding geometric formulas holding among prescribed geometric magnitudes which follow from the given assumptions.

Assume that the equations $h_1(x_1,\ldots,x_n) = 0,\ldots,h_r(x_1,\ldots,x_n) = 0$ express geometric properties of some objects. Let $x_1,\ldots,x_m$ be independent variables and $x_{m+1},\ldots,x_n$ be dependent variables. Eliminating variables (dependent or independent) we get the elimination ideal which contains only polynomials in those variables we did not eliminate. Usually we eliminate independent variables $x_1,\ldots,x_m$ or, if needed, some dependent variables $x_{m+1},\ldots,x_p$, $m \leq p \leq n$ to obtain a geometric statement expressed by the equation $c(x_{p+1},\ldots,x_n) = 0$ which follows from the assumptions $h_1,\ldots,h_r = 0$. The theorem holds:

Theorem 2.10. *Let* $I = (h_1,\ldots,h_r) \subset K[x_1,\ldots,x_n]$ *and* $c \in I \cap K[x_{p+1},\ldots,x_n]$, *for* $p \leq n$. *Then*

$$h_1(x_1,\ldots,x_n) = 0,\ldots,h_r(x_1,\ldots,x_n) = 0 \Rightarrow c(x_{p+1},\ldots,x_n) = 0. \quad (2.15)$$

Proof. The proof follows immediately from the inclusion $\{c\} \subset I$. Then by Theorem 2.8 on the ideal-variety correspondence $V(I) \subset V(c)$. This proves the theorem. □

We will show the method of automatic deriving in the following example.

Given a triangle ABC with side lengths a, b, c. Express the area p of a triangle ABC in terms of a, b, c.

Choose a Cartesian system of coordinates so that for the coordinates of the vertices A, B, C of a triangle $A = [0, 0]$, $B = [c, 0]$, $C = [x, y]$ (Fig. 2.5). We want to express the area p of a triangle ABC in terms of its side lengths $a = |BC|$, $b = |CA|$, $c = |AB|$. It is obvious that

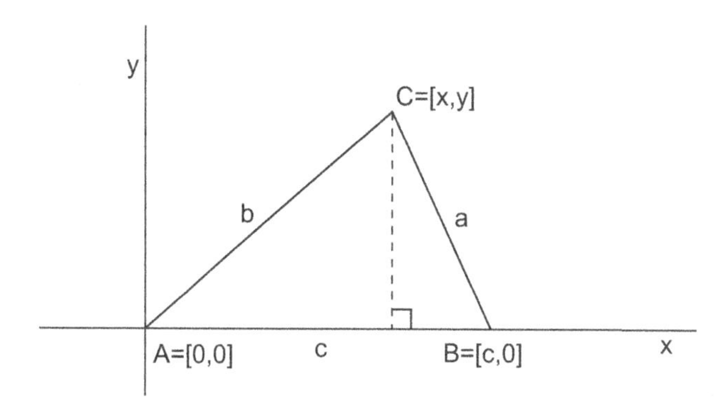

Fig. 2.5 Formula of Heron

$b = |AC| \Leftrightarrow h_1 : x^2 + y^2 - b^2 = 0,$

$a = |BC| \Leftrightarrow h_2 : (x - c)^2 + y^2 - a^2 = 0,$

$p = \text{area of } ABC \Leftrightarrow h_3 : p - \frac{1}{2}cy = 0.$

In the ring of polynomials $K[a, b, c, x, y, p]$ consider the ideal $I = (x^2 + y^2 - b^2, (x - c)^2 + y^2 - a^2, p - \frac{1}{2}cy)$. We search for such a formula which expresses the relation between the side lengths a, b, c of a triangle ABC and its area p. Such a polynomial belongs to the elimination ideal $I \cap K[a, b, c, p]$. Elimination of variables x, y in I

```
Use R::=Q[xyabcp];
I:=Ideal(x^2+y^2-b^2,(x-c)^2+y^2-a^2,2p-cy);
Elim(x..y,I);
```

gives the equation

$$16p^2 = -a^4 - b^4 - c^4 + 2a^2b^2 + 2a^2c^2 + 2b^2c^2, \tag{2.16}$$

which after a factorization has the form

$$16p^2 = (a + b + c)(-a + b + c)(a - b + c)(a + b - c). \tag{2.17}$$

We see that the last relation is the same as

$$p = \sqrt{s(s - a)(s - b)(s - c)}, \tag{2.18}$$

where $s = \frac{1}{2}(a+b+c)$.

We derived the well-known formula of Heron.

Let us give a classical approach of finding the formula of Heron.

The area of the triangle in Fig. 2.6 is given by

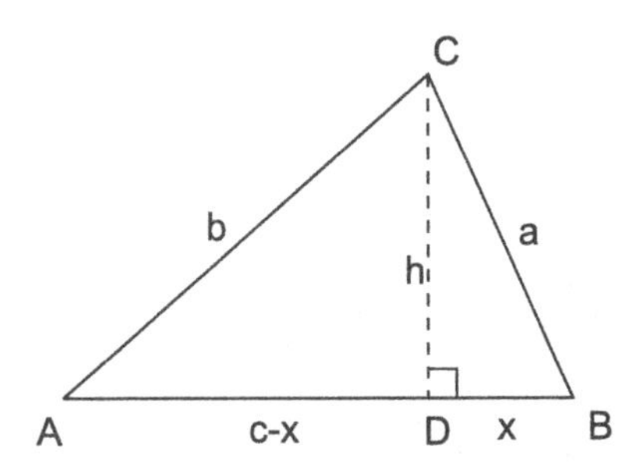

Fig. 2.6 Classical proof of Heron's formula

$$p = \frac{1}{2}ch. \tag{2.19}$$

Hence it is necessary to express the altitude h in terms of a, b, c. For the right triangles ADC and CDB the Pythagorean theorem relations

$$h^2 = b^2 - (c-x)^2, \quad h^2 = a^2 - x^2 \tag{2.20}$$

follow. This implies $b^2 - (c-x)^2 = a^2 - x^2$ with $2cx = a^2 - b^2 + c^2$. Substituting x into (2.20) we get

$$4c^2h^2 = 4a^2c^2 - (a^2 - b^2 + c^2)^2. \tag{2.21}$$

Finally a substitution for h from (2.21) into (2.19) gives

$$16p^2 = 4c^2h^2 = 4a^2c^2 - (a^2 - b^2 + c^2)^2 =$$
$$(2ac + a^2 - b^2 + c^2)(2ac - a^2 + b^2 - c^2) =$$
$$((a+c)^2 - b^2)(b^2 - (a-c)^2) = (a+c+b)(a+c-b)(b+a-c)(b-a+c),$$

which is the relation (2.17).

Remark 2.3. Both proofs — computer and classical ones — considerably differ. If the situation admits we could give a classical proof together with a computer one in school mathematics. Solving more complex problems would help students to realize that a classical method is often insufficient and that a computer method is more useful. However, it is noteworthy that both methods have their strengths and weaknesses.

2.5 Automatic discovery

By automatic discovery of theorems [102, 136] we mean the process of dealing automatically with arbitrary geometric statements (i.e. statements that could be, in general, false) and aiming to find complementary hypotheses such that the statements are true.

In the previous parts we have been concerned with a statement of the form

$$\forall x \in L^n, \quad h_1(x) = 0, \ldots, h_r(x) = 0 \quad \Rightarrow \quad c(x) = 0. \tag{2.22}$$

The statement (2.22) can be generally true or, after adding non-degeneracy conditions, generically true. However, it may happen that adding neither non-degeneracy condition helps the statement (2.22) become generically true. Then the situation in Fig. 2.7 occurs.

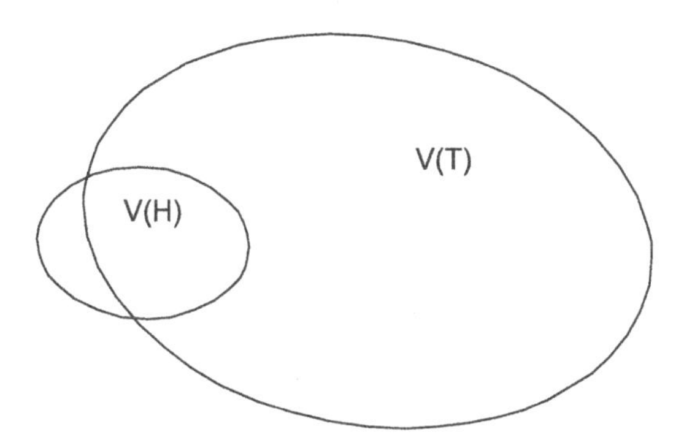

Fig. 2.7 The implication $H \Rightarrow T$ does not hold since $V(H) \not\subset V(T)$

We can react to this situation in two ways. In order for the statement (2.22) to be true we can

either

A) *"reduce"* the hypotheses variety $V(H)$

or

B) *"extend"* the conclusion variety $V(T)$.

In literature usually only the case A is mentioned. In practice we encounter the cases which can be solved by the method B, whereas the method A fails. First we will deal with the method A.

Reduction of the hypotheses variety $V(H)$ by the method A can be achieved by adding additional assumptions $h'_1, h'_2, \ldots, h'_p$ to the given assumptions $h_1, h_2, \ldots, h_r$ so that now the hypotheses variety is in the form of the *intersection* $V(H) \cap V(H')$. Obviously $V(H) \cap V(H') \subset V(H)$ holds. Our aim is to find such a set of additional assumptions H' so that $V(H) \cap V(H') \subset V(T)$. We proceed in the following manner.

We add a conclusion polynomial c to the ideal $I = (h_1, \ldots, h_r)$ and get the ideal $J = I \cup \{c\} = (h_1, \ldots, h_r, c)$. In the ideal J we eliminate *dependent* variables $x_{m+1}, \ldots, x_n$ and obtain the elimination ideal

$$(h_1, \ldots, h_r, c) \cap K[x_1, \ldots, x_m].$$

Denote by $h'_1, \ldots, h'_p$ polynomials which are involved in the elimination ideal. These polynomials could be the desired additional conditions which are necessary to add to the given assumptions so that the statement becomes valid. In this case we acquire a "smaller" hypotheses variety given by the ideal $(h_1, \ldots, h_r, h'_1, \ldots, h'_p)$ which is likely involved in the conclusion variety c. Now the statement has the form

$$\forall x \in L^n, \quad h_1(x) = 0, \ldots, h_r(x) = 0, h'_1(x) = 0, \ldots, h'_p(x) = 0 \Rightarrow c(x) = 0 \tag{2.23}$$

and the whole process of automatic theorem proving repeats with one change — instead of (2.22) we shall prove the statement (2.23).[3]

We explain the basic steps of automatic discovery theorems in the next example:

Statement 1. *Consider a triangle ABC and denote K, L, M the feet of perpendiculars dropped from an arbitrary point P to the sides BC, AC, AB of a triangle ABC respectively. Then the points K, L, M are collinear.*

To prove Statement 1 choose a Cartesian coordinate system so that $A = [a, 0]$, $B = [b, c]$, $C = [0, 0]$, $P = [p, q]$, $K = [m, n]$, $L = [k, 0]$, $M = [r, s]$ (see Fig. 2.8). Assumptions are as follows:

$PL \perp AC \Leftrightarrow h_1 : p - k = 0,$

$K \in BC \Leftrightarrow h_2 : cm - bn = 0,$

$PK \perp BC \Leftrightarrow h_3 : (p - m)b + (q - n)c = 0,$

$M \in AB \Leftrightarrow h_4 : (b - a)s - (r - a)c = 0,$

$PM \perp AB \Leftrightarrow h_5 : (p - r)(b - a) + (q - s)c = 0.$

[3]We warn the reader that this method eventually cannot conclude anything (see the paper [102], where a counterexample is provided).

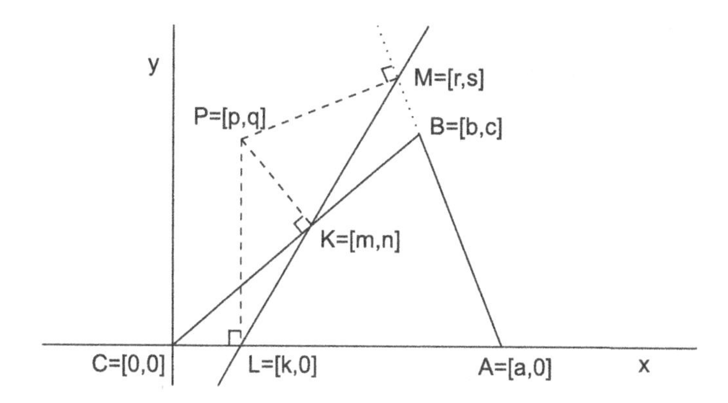

Fig. 2.8 Feet of perpendiculars K, L, M dropped from a point P to sides of ABC are collinear

The conclusion c of Statement 1 is expressed by the equation

K, L, M are collinear $\Leftrightarrow c : (m-k)s - (r-k)n = 0.$

In the ideal $I = (h_1, h_2, \ldots, h_5, c')$, where $c' = ((m-k)s - (r-k)n)t - 1$ is the negation of c with a slack variable t, we investigate the normal form of 1. We enter

```
Use R::=Q[abcpqkmnrst];
I:=Ideal(p-k,cm-bn,(p-m)b+(q-n)c,(b-a)s-(r-a)c,(b-a)(p-r)+
(q-s)c,((m-k)s-(r-k)n)t-1);
NF(1,I);
```

and obtain the answer 1. Hence the normal form is non-zero and Statement 1 is not generally true.

Now we search for non-degeneracy conditions. In the ideal I we eliminate a slack variable t and dependent variables k, m, n, r, s. We get

```
Use R::=Q[abcpqkmnrst];
I:=Ideal(p-k,cm-bn,(p-m)b+(q-n)c,(b-a)s-(r-a)c,(b-a)(p-r)+
(q-s)c,((m-k)s-(r-k)n)t-1);
Elim(k..t,I);
```

the answer `Ideal(0)` which does not give any non-degeneracy conditions.

In the next step we will try to find additional conditions which are necessary to add to assumptions so that Statement 1 becomes valid. We add the conclusion polynomial $(m-k)s - (r-k)n$ to the ideal I and in this new ideal $J = I \cup \{(m-k)s - (r-k)n\}$ we eliminate *dependent* variables k, m, n, r, s. The elimination ideal contains polynomials of independent variables a, b, c, p, q. CoCoA returns

```
Use R::=Q[abcpqkmnrs];
J:=Ideal(p-k,cm-bn,(p-m)b+(q-n)c,(b-a)s-(r-a)c,(b-a)(p-r)+
(q-s)c,(m-k)s-(r-k)n);
Elim(k..s,J);
```

the only polynomial of the form $ac^2(-acp + cp^2 + abq - b^2q - c^2q + cq^2)$. Relations $a = 0$ and $c = 0$ mean that the vertices of a triangle coincide or are collinear which may be ruled out. The remaining equation

$$-acp + cp^2 + abq - b^2q - c^2q + cq^2 = 0 \tag{2.24}$$

is a condition for a point P to lie on the circumcircle of the triangle ABC. This can be seen from the fact that considering points A, B, C as fixed and a point P as varying, then (2.24) is the equation of a circle which is circumscribed to the triangle ABC as we can easily find out by a direct substitution. The equation (2.24) could be the desired additional condition. We add the condition (2.24) to the assumptions of Statement 1. Now we will explore the following "newly" discovered Statement 2:

Statement 2. *Consider a triangle ABC and P an arbitrary point of the circumcircle of ABC. Denote by K, L, M the feet of perpendiculars dropped from P to the sides BC, AC, AB of a triangle respectively. Then the points K, L, M are collinear.*

Let us prove Statement 2.[4] We will investigate the normal form of the constant polynomial 1 with respect to the ideal K. We get

```
Use R::=Q[abcpqkmnrst];
K:=Ideal(p-k,cm-bn,(p-m)b+(q-n)c,(b-a)s-(r-a)c,(b-a)(p-r)+
(q-s)c,ac^2(-acp+cp^2+abq-b^2q-c^2q+cq^2),((m-k)s-(r-k)n)t-1);
NF(1,K);
```

the answer 1. Thus Statement 2 is not generally true.

Now we will explore non-degeneracy conditions. In the ideal K we eliminate a slack variable t and dependent variables q, k, m, n, r, s. It is worth noting that in this case, we consider the variable p as independent and q as dependent for it is constrained by the equation of a circle (2.24). Sometimes we call such variables p, q as *semidependent.* We enter

[4] The same example of discovery already appeared in the book by T. Recio [103], which provides the largest collection of examples of automatic discovery up to date.

```
Use R::=Q[abcpqkmnrst];
K:=Ideal(p-k,cm-bn,(p-m)b+(q-n)c,(b-a)s-(r-a)c,(b-a)(p-r)+
(q-s)c,ac^2(-acp+cp^2+abq-b^2q-c^2q+cq^2),((m-k)s-(r-k)n)t-1);
Elim(q..t,K);
```

and get one polynomial which leads to the equation

$$(b^2 + c^2)(a^2 - 2ab + b^2 + c^2) = 0. \tag{2.25}$$

First, we will find out what this condition means geometrically. We rearrange the condition (2.25) into the form

$$(b^2 + c^2)((a - b)^2 + c^2) = 0, \tag{2.26}$$

from which we see that for the vertices of a triangle either $A = C$ (which is expressed by the equality $b^2 + c^2 = 0$) or $B = C$ (which is given by $(a - b)^2 + c^2 = 0$) hold. We need to rule out these cases of degeneracy. Let us suppose that $A \neq C$ and $B \neq C$ and add the negation of (2.26)

$$(b^2 + c^2)((a - b)^2 + c^2)v - 1 = 0, \tag{2.27}$$

where v is a slack variable, to the ideal K. We get

```
Use R::=Q[abcpqkmnrsvt];
L:=Ideal(p-k,cm-bn,(p-m)b+(q-n)c,(b-a)s-(r-a)c,(b-a)(p-r)+
(q-s)c,ac^2(-acp+cp^2+abq-b^2q-c^2q+cq^2),(b^2+c^2)((a-b)^2+
c^2)v-1,((m-k)s-(r-k)n)t-1);
NF(1,L);
```

the result 0 which means that Statement 2 is generically true.

Remark 2.4.

1) We arrived at Statement 2 adding the additional condition "a point P lies on the circumcircle of a triangle ABC" to Statement 1. Many readers would doubtlessly recognize in Statement 2 the well-known Simson–Wallace theorem [26] which was in this way rediscovered (Fig. 2.9, see Chapter 4).

2) When solving problems automatically we first translated a given geometric situation from geometry to algebra to obtain the system of equations and inequations. Then we solved the given problem in algebra using the theory of elimination. During this process and at the end of it we have to translate found results back from algebra to geometry. This part is very difficult especially for students (see for example the relations (2.24) and (2.25)). Translation from algebra back into geometry is, in general, not easy.

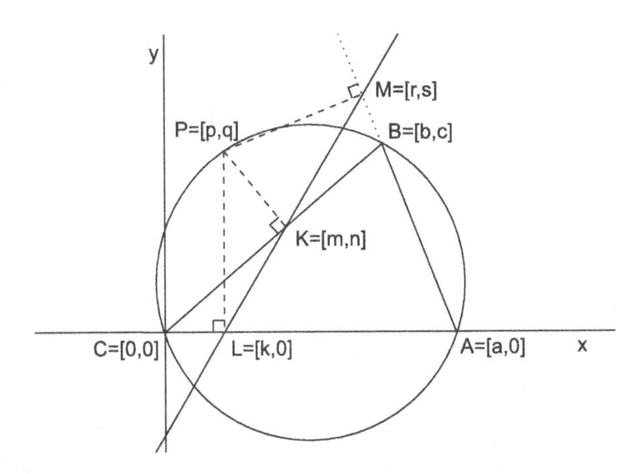

Fig. 2.9 Feet of perpendiculars K, L, M dropped from a point P of the circumcircle to the sides of a triangle ABC are collinear

Now we return to the beginning of Section 2.5 "Automatic discovery". We will focus on the second method of investigation of the statement

$$\forall x \in L^n, \quad h_1(x) = 0, \ldots, h_r(x) = 0 \quad \Rightarrow \quad c(x) = 0 \tag{2.28}$$

which is, in general, not true. We will use the method B which is based on *extension* of the conclusion variety $V(T)$, unlike the method A which is based on *reduction* of the hypotheses variety $V(H)$.

In order to extend the conclusion variety $V(T)$ we add additional conditions to the conclusion $c = 0$. If these additional conditions are expressed by the equations $c'_1 = 0, \ldots, c'_p = 0$, then a new conclusion variety has the form of the *union* $V(T) \cup V(T')$, where $V(T') = \{c'_1 = 0\} \cup \cdots \cup \{c'_p = 0\}$. Obviously $V(T) \subset V(T) \cup V(T')$, hence this new conclusion variety is "greater". We are to find such a set of additional conditions T' so that after adding them to the set of conclusions T the relation $V(H) \subset V(T) \cup V(T')$ is fulfilled. The question is how to find these additional conditions. We will proceed in the following way:

To the ideal $I = (h_1, \ldots, h_r)$ we *add* negation $ct - 1$ of the conclusion polynomial c, where t is a slack variable. We get the ideal

$$J = I \cup \{ct - 1\} = (h_1, \ldots, h_r, ct - 1).$$

In the ideal J we eliminate *independent* variables $x_1, \ldots, x_m$ plus a slack variable t to obtain the elimination ideal

$$(h_1, \ldots, h_r, ct - 1) \cap K[x_{m+1}, \ldots, x_n]$$

in which additional conditions $c'_1, \ldots, c'_p$ will be likely involved.

We add these polynomials to the conclusion polynomial c and form a *product* $c \cdot c'_1 \cdot \ldots \cdot c'_p$. Instead of (2.28) we get a new statement of the form

$$\forall x \in L^n, \quad h_1(x) = 0, \ldots, h_r(x) = 0 \quad \Rightarrow \quad c(x) \cdot c'_1(x) \cdot \ldots \cdot c'_p(x) = 0. \tag{2.29}$$

Now the whole procedure will repeat with one change — instead of the statement (2.28) we will prove the statement (2.29). Let us see the next example to explain a matter of automatic discovery by the method B:

Statement 3: *Given a cyclic quadrilateral with side lengths a, b, c, d and diagonals e, f. Then $ac + bd = ef$.*

Statement 3 is known from school geometry as the Ptolemy's theorem. Let us try to prove it.

Suppose that $ABCD$ is a quadrilateral with sides $a = |AB|$, $b = |BC|$, $c = |CD|$, $d = |AD|$ and diagonals $e = |AC|$, $f = |BD|$, whose vertices lie on the circumcircle with the center S and the radius r.

Choose a Cartesian coordinate system so that $A = [0,0]$, $B = [a,0]$, $C = [x,y]$, $D = [u,v]$, $S = [s,t]$ (Fig. 2.10). The hypotheses are as follows:

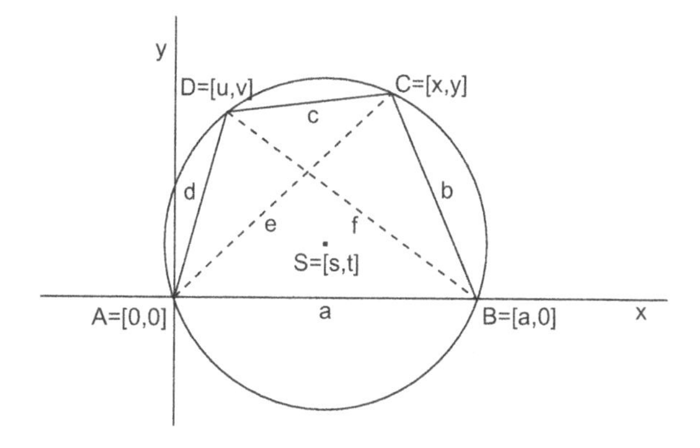

Fig. 2.10 Ptolemy's theorem: $ac + bd = ef$

$r = |AS| \Leftrightarrow h_1 : s^2 + t^2 - r^2 = 0,$
$r = |BS| \Leftrightarrow h_2 : (a - s)^2 + t^2 - r^2 = 0,$
$r = |CS| \Leftrightarrow h_3 : (x - s)^2 + (y - t)^2 - r^2 = 0,$
$r = |DS| \Leftrightarrow h_4 : (u - s)^2 + (v - t)^2 - r^2 = 0,$
$b = |BC| \Leftrightarrow h_5 : (a - x)^2 + y^2 - b^2 = 0,$

$c = |CD| \Leftrightarrow h_6 : (u - x)^2 + (v - y)^2 - c^2 = 0,$
$d = |DA| \Leftrightarrow h_7 : u^2 + v^2 - d^2 = 0,$
$e = |AC| \Leftrightarrow h_8 : x^2 + y^2 - e^2 = 0,$
$f = |BD| \Leftrightarrow h_9 : (u - a)^2 + v^2 - f^2 = 0.$

The conclusion z has the form:

$z : ac + bd - ef = 0.$

Denoting $I = (h_1, h_2, \ldots, h_9)$ we investigate the normal form of 1 in the ideal $J = I \cup \{zk - 1\} = (h_1, h_2, \ldots, h_9, zk - 1)$, where k is a slack variable.

```
Use R::=Q[abcdefxyuvstrk];
J:=Ideal(s^2+t^2-r^2,(a-s)^2+t^2-r^2,(x-s)^2+(y-t)^2-r^2,
(u-s)^2+(v-t)^2-r^2,(a-x)^2+y^2-b^2,(u-x)^2+(v-y)^2-c^2,u^2+
v^2-d^2,x^2+y^2-e^2,(u-a)^2+v^2-f^2,(ac+bd-ef)k-1);
NF(1,J);
```

The result is 1, hence Statement 3 is not generally true.

Now we shall explore non-degeneracy conditions. In the ideal J we eliminate a slack variable k and dependent variables b, c, d, e, f, s, t, r.

```
Use R::=Q[abcdefstrkxyuv];
J:=Ideal(s^2+t^2-r^2,(a-s)^2+t^2-r^2,(x-s)^2+(y-t)^2-r^2,
(u-s)^2+(v-t)^2-r^2,(a-x)^2+y^2-b^2,(u-x)^2+(v-y)^2-c^2,u^2+
v^2-d^2,x^2+y^2-e^2,(u-a)^2+v^2-f^2,(ac+bd-ef)k-1);
Elim(b..k,J);
```

We get the only equation

$$-ayu + yu^2 + axv - x^2v - y^2v + yv^2 = 0 \tag{2.30}$$

which expresses a circle circumscribed to a quadrilateral and does not give any non-degeneracy condition.

Searching for additional conditions which are necessary to add to given assumptions so that Statement 3 becomes valid by the method A gives

```
Use R::=Q[abcdefstrxyuv];
J:=Ideal(s^2+t^2-r^2,(a-s)^2+t^2-r^2,(x-s)^2+(y-t)^2-r^2,
(u-s)^2+(v-t)^2-r^2,(a-x)^2+y^2-b^2,(u-x)^2+(v-y)^2-c^2,u^2+
v^2-d^2,x^2+y^2-e^2,(u-a)^2+v^2-f^2,ac+bd-ef);
Elim(b..r,J);
```

the condition (2.30) that is of no help.

Now we will use the method B, i.e., we search for additional conditions which are necessary to add to the conclusion z so that the statement becomes valid. This means that in the ideal

$$J = I \cup \{(ac + bd - ef)k - 1\}$$

we will eliminate a slack variable k and *independent* variables x, y, u, v, s, t, r. We get

```
Use R::=Q[abcdefxyuvstrk];
J:=Ideal(s^2+t^2-r^2,(a-s)^2+t^2-r^2,(x-s)^2+(y-t)^2-r^2,
(u-s)^2+(v-t)^2-r^2,(a-x)^2+y^2-b^2,(u-x)^2+(v-y)^2-c^2,u^2+
v^2-d^2,x^2+y^2-e^2,(u-a)^2+v^2-f^2,(ac+bd-ef)k-1);
Elim(x..k,J);
```

the elimination ideal which contains 13 polynomials. One of them has the form $(ac - bd - ef)(ac - bd + ef)(ac + bd + ef)$. This could be the desired condition. Adding this polynomial to the conclusion polynomial $ac+bd-ef$, (i.e., as a product) we check the statement again.

```
Use R::=Q[abcdefxyuvstrk];
K:=Ideal(s^2+t^2-r^2,(a-s)^2+t^2-r^2,(x-s)^2+(y-t)^2-r^2,
(u-s)^2+(v-t)^2-r^2,(a-x)^2+y^2-b^2,(u-x)^2+(v-y)^2-c^2,u^2+
v^2-d^2,x^2+y^2-e^2,(u-a)^2+v^2-f^2,
(ac+bd-ef)(ac-bd-ef)(ac-bd+ef)(ac+bd+ef)k-1);
NF(1,K);
```

The result 0 means that this new statement is generally true.

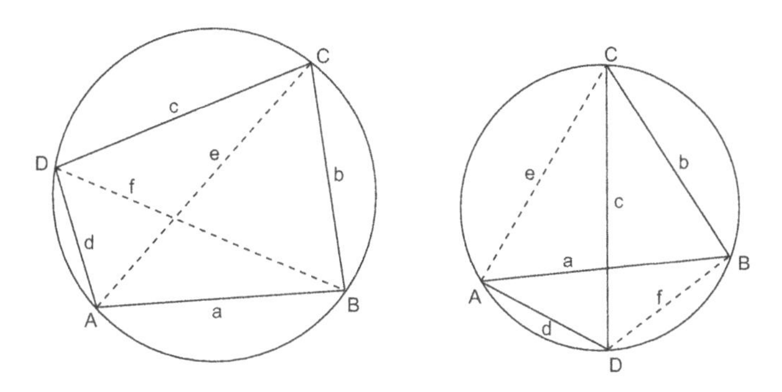

Fig. 2.11 Cyclic quadrilaterals with sides a, b, c, d — convex and non-convex cases

We obtained the following Statement 4:

Statement 4: *Given a cyclic quadrilateral with side lengths a, b, c, d and diagonals e, f. Then*

$$(ac + bd - ef)(ac - bd - ef)(ac - bd + ef)(ac + bd + ef) = 0.$$

What is the geometric meaning of the added conditions? A cyclic quadrilateral with the given side lengths and diagonals may have besides the form in Fig. 2.11 on the left another form as well (see Fig. 2.11 on the right). It depends on the position of the vertex D on the circumcircle of ABC with respect to the vertices A, B, C. Thus we rediscovered the Ptolemy's theorem (see the relation (3.24) in Chapter 3 "Generalization of the formula of Heron").

Chapter 3

Generalization of the formula of Heron

Every student knows the formula of Heron for the area p of a triangle with side lengths a, b, c

$$p = \sqrt{s(s-a)(s-b)(s-c)}, \tag{3.1}$$

where $s = 1/2(a+b+c)$. This formula is usually ascribed to Heron of Alexandria c. 60 B.C., although it was likely known to Archimedes, 287–212 B.C.

The formula of Brahmagupta (Brahmagupta — Indian mathematician, 598–c. 665 A.D.) for the area p of a convex cyclic quadrilateral which is given by the side lengths a, b, c, d is as follows:

$$p = \sqrt{(s-a)(s-b)(s-c)(s-d)}, \tag{3.2}$$

where $s = 1/2(a+b+c+d)$, is a generalization of the formula of Heron.

Since that time, despite a great effort of mathematicians from all over the world, no similar formula for the area of a cyclic n- gon for $n > 4$ has appeared until 1994 when American D. P. Robbins published [112]. Almost 1400 years the formula for the area of a cyclic pentagon was missing. In the meantime some works on computation of the area of cyclic polygons of special classes appeared (see for example [5, 26, 118]). The main reason for the long time elapse is a big complexity of such formulas.

In this chapter we will find a formula for the area of a cyclic pentagon by computer methods which are based on results of computational commutative algebra as Gröbner bases of ideals, elimination of variables and the theory of automatic theorem proving and discovering. It seems that without using these methods we would hardly arrive at these results.

We will start with well-known instances of a triangle and a quadrilateral and then go on to investigate a cyclic pentagon [92]. See also the last results [38, 71, 85, 129].

3.1 Area of a polygon

In this section we will be concerned with the area of planar polygons.

Let us denote by $A_1A_2\dots A_n$ a *closed polygon* with vertices $A_1, A_2, \dots, A_n$. All indices are considered mod n, i.e., $A_{j+n} = A_j$ for all $j = 1, 2, \dots, n$. Instead of a closed n-gon we will say merely an n-gon because we will not use other types in this text. If it is not necessary to stress the number of vertices we simply talk about a polygon. Under the word area we will always understand the *signed* area — hence the area may even be a negative number.

Computation of the area of a polygon will be carried out in two basic ways.

The first way consists of computing the area of a polygon once knowing the coordinates of its vertices in a given system of coordinates. Then the area of a polygon can be computed by the following theorem (see W. Blaschke [8]):

Theorem 3.1. *Let $A_i = [x_i, y_i]$, $i = 1, 2, \dots, n$, be coordinates of the vertices of an n-gon $A_1A_2\dots A_n$ in a given Cartesian system of coordinates. Then for the area p of an n-gon $A_1A_2\dots A_n$*

$$p = \frac{1}{2}\sum_{i=1}^{n}\begin{vmatrix} x_i & y_i \\ x_{i+1} & y_{i+1} \end{vmatrix}. \tag{3.3}$$

We can easily check that the formula (3.3) does not depend on a choice of the system of coordinates. By (3.3) we compute the area of an n-gon as the sum of (signed) areas of individual triangles, which the given n-gon consists of.

The second way of computing the area of a polygon is based on distances between the vertices of a polygon. The area p of an n-gon $A_1A_2\dots A_n$ can be expressed in terms of all $\binom{n}{2}$ mutual distances between its vertices. We will use the formula (3.4) which was published in 1949 by B. Sz. Nagy and L. Rédey [105]. The theorem reads:

Theorem 3.2 (Nagy–Rédey). *Let $d_{ij} = |A_iA_j|^2$ denote a square of the distance of the vertices A_i, A_j. Then the area p of an n-gon $A_1A_2\dots A_n$ is given by*

$$16p^2 = \sum_{i,j=1}^{n}\begin{vmatrix} d_{i,j} & d_{i,j+1} \\ d_{i+1,j} & d_{i+1,j+1} \end{vmatrix}. \tag{3.4}$$

We derive the formula (3.4) for some n both by computer and in a classical way. The formula of Heron for the area of a triangle was derived in the previous chapter. Now we will be concerned with a less known formula for the area of a quadrilateral — the formula of Staudt.

3.1.1 *Formula of Staudt*

It is known that an arbitrary quadrilateral is not determined by its four side lengths. To determine it, we need to add another condition, for instance, the length of one diagonal. But neither of these five conditions (four sides plus a diagonal) determine a quadrilateral uniquely. We will explore the area of a quadrilateral knowing all six distances between its vertices.

Let $ABCD$ be a quadrilateral in the plane with side lengths a, b, c, d and diagonals e, f. Express the area p of a quadrilateral $ABCD$ in terms of a, b, c, d, e, f.

Denote $a = |AB|$, $b = |BC|$, $c = |CD|$, $d = |DA|$, $e = |AC|$, $f = |BD|$. Choose a Cartesian system of coordinates so that for the vertices of a quadrilateral $A = [0, 0]$, $B = [a, 0]$, $C = [x, y]$, $D = [u, v]$ (Fig. 3.1). We have the following relations:

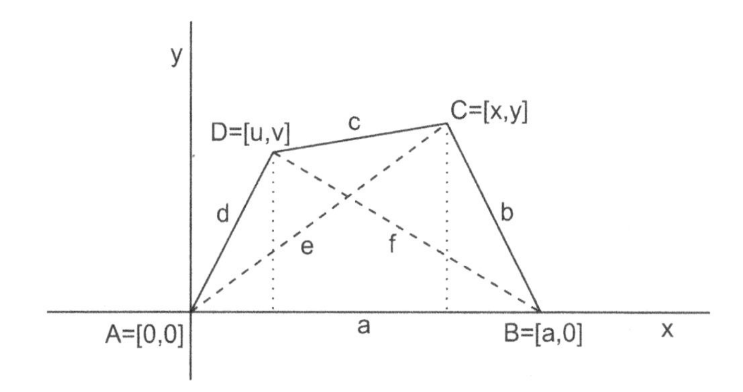

Fig. 3.1 Deriving the formula of Staudt

$$
\begin{aligned}
b = |BC| &\Leftrightarrow h_1 : (x - a)^2 + y^2 - b^2 = 0, \\
c = |CD| &\Leftrightarrow h_2 : (u - x)^2 + (v - y)^2 - c^2 = 0, \\
d = |DA| &\Leftrightarrow h_3 : u^2 + v^2 - d^2 = 0, \\
e = |AC| &\Leftrightarrow h_4 : x^2 + y^2 - e^2 = 0,
\end{aligned}
$$

$f = |BD| \Leftrightarrow h_5 : (u-a)^2 + v^2 - f^2 = 0.$

If we drop perpendiculars from the vertices C and D to the side AB we divide a quadrilateral $ABCD$ into two right triangles and one trapezoid, whose sum of areas equals the area p of $ABCD$ (Fig. 3.1). After a short computation we obtain

$p =$ area of $ABCD \Leftrightarrow h_6 : p - 1/2(ay + xv - yu) = 0.$

We obtain the same result using the formula (3.3).
In the ideal $I = (h_1, h_2, \ldots, h_6)$ we eliminate variables x, y, u, v. We enter

```
Use R::=Q[xyuvpabcdef];
I:=Ideal((x-a)^2+y^2-b^2,(u-x)^2+(v-y)^2-c^2,u^2+v^2-d^2,x^2
+y^2-e^2,(u-a)^2+v^2-f^2,p-1/2(ay+xv-yu));
Elim(x..v,I);
```

and get four polynomials from which the following

$$16p^2 = 4e^2f^2 - (a^2 - b^2 + c^2 - d^2)^2 \tag{3.5}$$

is of our interest. We derived the formula (3.5) which is a special case of (3.4). It was published in 1842 by Ch. R. Staudt [128]. That is why we call (3.5) the formula of Staudt.

Remark 3.1.
1) Note that setting $d = 0$ into (3.5), we get the formula of Heron (2.16).

2) The formula (3.5) is valid for any choice of the vertices A, B, C, D of a quadrilateral, similarly for the case when a quadrilateral is non-convex or even intersects itself.

Instead of eliminating x, y, u, v in the ideal I we can eliminate variables x, y, u, v, p as well. This elimination leads to the so-called *Euler's four points relation* (see [29]):

$a^4c^2 - a^2b^2c^2 + a^2c^4 - a^2b^2d^2 + b^4d^2 - a^2c^2d^2 - b^2c^2d^2 + b^2d^4 + a^2b^2e^2 - a^2c^2e^2 - b^2d^2e^2 + c^2d^2e^2 - a^2c^2f^2 + b^2c^2f^2 + a^2d^2f^2 - b^2d^2f^2 - a^2e^2f^2 - b^2e^2f^2 - c^2e^2f^2 - d^2e^2f^2 + e^4f^2 + e^2f^4 = 0.$

This relation expresses mutual dependence of all six distances a, b, c, d, e, f between four vertices of a quadrilateral. The Euler's four points relation follows from the Cayley–Menger determinant [5], for the volume V of a tetrahedron with the edges of lengths a, b, c, d, e, f

$$288V^2 = \begin{vmatrix} 0 & 1 & 1 & 1 & 1 \\ 1 & 0 & a^2 & e^2 & d^2 \\ 1 & a^2 & 0 & b^2 & f^2 \\ 1 & e^2 & b^2 & 0 & c^2 \\ 1 & d^2 & f^2 & c^2 & 0 \end{vmatrix}, \tag{3.6}$$

if we set $V = 0$. A comparison of the equality $V = 0$ from (3.6) with the Euler's four points relation shows that both relations are alike up to the constant factor -2 (cf. (3.6) with the formula of Ptolemy (3.25)).

To compare computer and classical approaches, we will give a classical proof of the formula of Staudt (3.5).

This will require some "idea". From the right triangles AED and DEC,

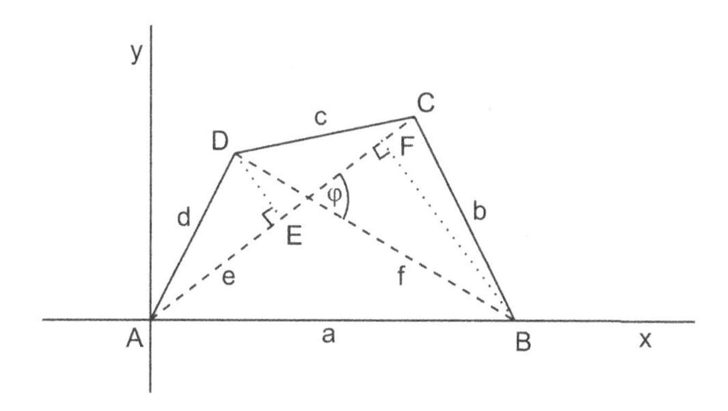

Fig. 3.2 A classical proof of the formula of Staudt

Fig. 3.2 relations

$$|DE|^2 = d^2 - |AE|^2, \; |DE|^2 = c^2 - |EC|^2$$

follow. The equality of the left sides implies the equality of the right sides

$$d^2 - |AE|^2 = c^2 - |EC|^2. \tag{3.7}$$

Analogously from the right triangles AFB and CFB

$$a^2 - |AF|^2 = b^2 - |FC|^2 \tag{3.8}$$

follows. By adding the equalities (3.7) and (3.8) we obtain

$$a^2 - b^2 + c^2 - d^2 = |AF|^2 - |FC|^2 + |EC|^2 - |AE|^2. \tag{3.9}$$

The right-hand side in (3.9) can be written in the form

$|AF|^2 - |FC|^2 + |EC|^2 - |AE|^2 =$

$(|AF| - |FC|)(|AF| + |FC|) + (|EC| - |AE|)(|EC| + |AE|) =$

$e(|AF| - |FC| + |EC| - |AE|) = \pm 2e|EF|,$

i.e.

$$(a^2 - b^2 + c^2 - d^2)^2 = 4e^2|EF|^2. \tag{3.10}$$

Further we see that $|EF| = f\cos\varphi$. Substitution into (3.10) gives

$$(a^2 - b^2 + c^2 - d^2)^2 = 4e^2 f^2 \cos^2\varphi. \tag{3.11}$$

Now we apply the well-known formula [4] for the area of a quadrilateral in terms of its lengths of diagonals e, f and the angle φ between diagonals

$$p = \frac{1}{2}\, ef \sin\varphi. \tag{3.12}$$

Substituting (3.12) into (3.11) with the use of $\sin^2\varphi = 1 - \cos^2\varphi$ gives the formula (3.5).

3.1.2 *Area of a pentagon and hexagon*

Let us derive by computer the Nagy–Rédey formula (3.4) for $n = 5$ and $n = 6$.

Given a pentagon $ABCDE$ in a plane, denote the sides and diagonals by $a = |AB|$, $b = |BC|$, $c = |CD|$, $d = |DE|$, $e = |EA|$, $i_1 = |AC|$, $i_2 = |AD|$, $i_3 = |BD|$, $i_4 = |BE|$, $i_5 = |CE|$.
Choose a Cartesian system of coordinates so that $A = [0,0]$, $B = [a,0]$, $C = [x,y]$, $D = [u,v]$, $E = [w,z]$ (Fig. 3.3). Then the following relations hold:

$|BC| = b \Leftrightarrow h_1 : (x-a)^2 + y^2 - b^2 = 0,$
$|CD| = c \Leftrightarrow h_2 : (u-x)^2 + (v-y)^2 - c^2 = 0,$
$|DE| = d \Leftrightarrow h_3 : (w-u)^2 + (z-v)^2 - d^2 = 0,$
$|EA| = e \Leftrightarrow h_4 : w^2 + z^2 - e^2 = 0,$
$|AC| = i_1 \Leftrightarrow h_5 : x^2 + y^2 - i_1^2 = 0,$
$|AD| = i_2 \Leftrightarrow h_6 : u^2 + v^2 - i_2^2 = 0,$
$|BD| = i_3 \Leftrightarrow h_7 : (u-a)^2 + v^2 - i_3^2 = 0,$
$|BE| = i_4 \Leftrightarrow h_8 : (w-a)^2 + z^2 - i_4^2 = 0,$
$|CE| = i_5 \Leftrightarrow h_9 : (x-w)^2 + (y-z)^2 - i_5^2 = 0.$

Area of $ABCDE = p \Leftrightarrow h_{10} : p - 1/2(ay + xv - yu + uz - vw) = 0.$

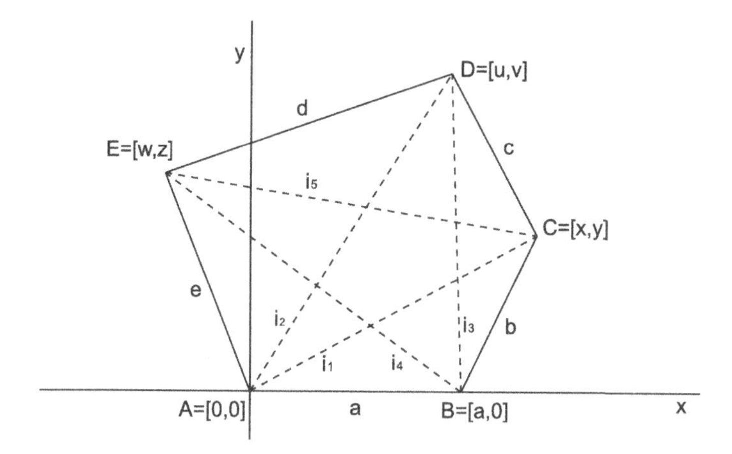

Fig. 3.3 Area of a pentagon $ABCDE$

Elimination of x, y, u, v, w, z in the ideal $I = (h_1, h_2, \ldots, h_{10})$ takes up CoCoA 9 minutes 18 seconds and it appears that

$$16p^2 = -(a^4 + b^4 + c^4 + d^4 + e^4) + 2(a^2b^2 + b^2c^2 + c^2d^2 + d^2e^2 + e^2a^2) \\ +2(i_1^2i_3^2 + i_2^2i_4^2 + i_3^2i_5^2 + i_4^2i_1^2 + i_5^2i_2^2) - 2(a^2i_5^2 + b^2i_2^2 + c^2i_4^2 + d^2i_1^2 \\ +e^2i_3^2) \quad (3.13)$$

which is the Nagy–Rédey formula (3.4) for $n = 5$.

In a similar way we get the following formula for the area of a hexagon. Denote

$a = |AB|$, $b = |BC|$, $c = |CD|$, $d = |DE|$, $e = |EF|$, $f = |FA|$, $i_1 = |AC|$, $i_2 = |AD|$, $i_3 = |AE|$, $i_4 = |BD|$, $i_5 = |BE|$, $i_6 = |BF|$, $i_7 = |CE|$, $i_8 = |CF|$, $i_9 = |DF|$ (Fig. 3.4).

Choosing a Cartesian system of coordinates so that $A = [0, 0]$, $B = [a, 0]$, $C = [x, y]$, $D = [u, v]$, $E = [w, z]$, $F = [s, t]$ we express all distances b, c, d, e, f, i_1, i_2, i_3, i_4, i_5, i_6, i_7, i_8, i_9 in terms of the coordinates of the vertices A, B, C, D, E, F of a hexagon similarly as in the case of the pentagon above. For the area p of $ABCDEF$ we have

$p - 1/2(ay + xv - yu + uz - vw + wt - zs) = 0.$

Eliminating variables x, y, u, v, w, z, s, t in the respective ideal we get in

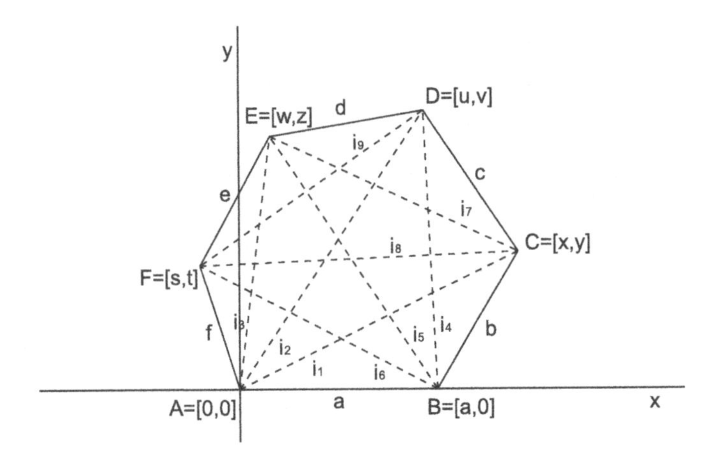

Fig. 3.4 Area of a hexagon

Singular[1] in 29 minutes the formula (together with 121 further polynomials)

$$16p^2 = -(a^4+b^4+c^4+d^4+e^4+f^4) + 2(a^2b^2+b^2c^2+c^2d^2+d^2e^2+e^2f^2 + f^2a^2) + 2(i_1^2i_4^2 - i_3^2i_4^2 + i_2^2i_5^2 + i_1^2i_6^2 + i_3^2i_6^2 + i_4^2i_7^2 - i_6^2i_7^2 + i_2^2i_8^2 + i_5^2i_8^2 - i_1^2i_9^2 + i_3^2i_9^2 + i_7^2i_9^2) - 2(a^2i_8^2 + b^2i_2^2 + c^2i_5^2 + d^2i_8^2 + e^2i_2^2 + f^2i_5^2), \qquad (3.14)$$

which is a special case of (3.4) for $n = 6$.

3.2 Area of a cyclic polygon

In this section we will explore properties of such polygons whose vertices lie on a circle. We call such polygons *cyclic* or *inscribed.* We will examine the properties of cyclic polygons in two ways, similarly as what we did with the area of a polygon.

Basically we can proceed in a *coordinate* way to introduce a Cartesian system of coordinates and express in a way such that the vertices of a cyclic n-gon lie on a circle etc. using the formula (3.3) for the area of an n-gon.

The other method, say *distance* or *coordinate-free* method, is based on the Nagy–Rédey formula (3.4) and Ptolemy's type conditions in which we use mutual distances between the vertices of a cyclic polygon. We shall see that the latter method is more effective.

[1]Software is available at http://www.singular.uni-kl.de

First we will derive a necessary and sufficient condition for the vertices of a quadrilateral to lie on a circle with known coordinates of these vertices. Then we concentrate on the well-known formula of Ptolemy which expresses a necessary and sufficient condition for a quadrilateral to be cyclic in terms of all distances between the vertices of a quadrilateral.

We will use the results of the previous sections to express the area of a cyclic n-gon for $n = 3, 4, 5$ which is given by its side lengths. For $n = 3$ we obtain the formula of Heron while for $n = 4$ we derive the well-known formula of Brahmagupta for the area of a convex cyclic quadrilateral and less known analogue of the Brahmagupta's formula for non-convex cyclic quadrilaterals.

First we will deal with cyclic quadrilaterals to show the essence of the technique. In the end we will derive the formula for the area of a cyclic pentagon [112]. We will find this formula using both methods — coordinate and distance. It turns out that without the use of modern computer algebra tools, solving this problem would be hardly feasible.

3.2.1 *Formula of Ptolemy*

About an arbitrary triangle with the side lengths a, b, c we can circumscribe a circle. Whence every triangle is cyclic. Besides the triangle inequality, no other constraint on a, b, c is necessary. Now we will pay attention to the case of a cyclic quadrilateral. We will be concerned with the following problem:

What is a necessary and sufficient condition for a quadrilateral $ABCD$ to be cyclic?

Suppose that $ABCD$ is a quadrilateral with sides and diagonals of lengths $a = |AB|$, $b = |BC|$, $c = |CD|$, $d = |AD|$, $e = |AC|$, $f = |BD|$ and whose vertices lie on a circle centered at the point S with the radius r.

First we will solve this problem by the coordinate method. We derive a necessary and sufficient condition for the coordinates of the vertices A, B, C, D.

Choose a Cartesian system of coordinates so that $A = [0, 0]$, $B = [a, 0]$, $C = [x, y]$, $D = [u, v]$, $S = [s, t]$ (Fig. 3.5). Then the following relations hold:

$r = |AS| \Leftrightarrow h_1 : s^2 + t^2 - r^2 = 0,$
$r = |BS| \Leftrightarrow h_2 : (a - s)^2 + t^2 - r^2 = 0,$
$r = |CS| \Leftrightarrow h_3 : (x - s)^2 + (y - t)^2 - r^2 = 0,$

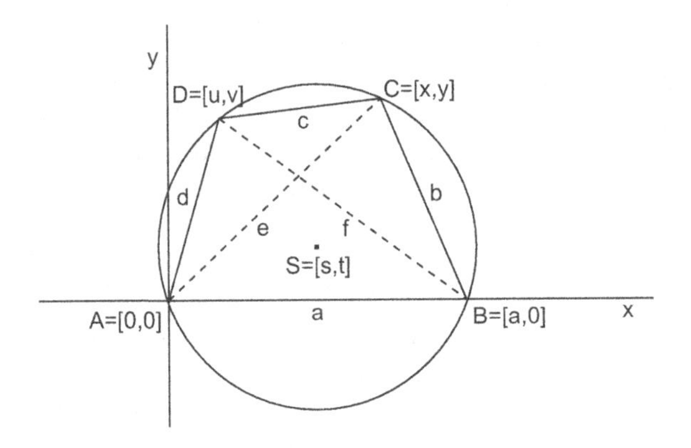

Fig. 3.5 Cyclic quadrilateral

$r = |DS| \Leftrightarrow h_4 : (u-s)^2 + (v-t)^2 - r^2 = 0.$

In the ideal $I = (h_1, h_2, h_3, h_4)$ we eliminate all variables besides a, x, y, u, v

```
Use R::=Q[axyuvstr];
I:=Ideal(s^2+t^2-r^2,(a-s)^2+t^2-r^2,(x-s)^2+(y-t)^2-r^2,
(u-s)^2+(v-t)^2-r^2);
Elim(s..r,I);
```

and obtain the condition $a(-ayu+yu^2+axv-x^2v-y^2v+yv^2) = 0$. Suppose that $a \neq 0$, i.e., the vertices A and B of a quadrilateral are distinct. Then

$$-ayu + yu^2 + axv - x^2v - y^2v + yv^2 = 0 \tag{3.15}$$

is a necessary condition for the vertices of $ABCD$ to lie on a circle.

Conversely we will prove that the condition (3.15) is sufficient as well. Assume that the vertices A, B, C lie on a circle with the center S and radius r and the condition (3.15) holds. We want to prove that the vertex D is on the same circle. We suppose that $a \neq 0$ which is expressed by $am - 1 = 0$, where m is a slack variable. Further realize that the equation $((u-s)^2 + (v-t)^2 - r^2)n - 1 = 0$, where n is another slack variable, means a negation of the conclusion of a statement.

```
Use R::=Q[axyuvstrmn];
J:=Ideal(s^2+t^2-r^2,(a-s)^2+t^2-r^2,(x-s)^2+(y-t)^2-r^2,am-1,
-ayu+ yu^2+axv-x^2v-y^2v+yv^2,((u-s)^2+(v-t)^2-r^2)n-1);
NF(1,J);
```

The response 1 means that the statement is not generally true. By elimination of independent variables s, t, r, m, n in the ideal J we find out non-degeneracy conditions which are necessary to rule out. We enter

```
Use R::=Q[axyuvstrmn];
J:=Ideal(s^2+t^2-r^2,(a-s)^2+t^2-r^2,(x-s)^2+(y-t)^2-r^2,am-1,
-ayu+yu^2+axv-x^2v-y^2v+yv^2,((u-s)^2+(v-t)^2-r^2)n-1);
Elim(s..n,J);
```

and get an answer

```
Ideal(y,ax - x^2)
```

which means that the reason of the problem could be hidden in the validity of equations $y = 0$ or $ax - x^2 = 0$. First we exclude the case $y = 0$, that is, we suppose that points A, B, C are not collinear. This could be written by the only equation $aym - 1 = 0$ (together with the condition $A \neq B$).

```
Use R::=Q[axyuvstrmn];
K:=Ideal(s^2+t^2-r^2,(a-s)^2+t^2-r^2,(x-s)^2+(y-t)^2-r^2,aym-1,
-ayu+yu^2+axv-x^2v-y^2v+yv^2,((u-s)^2+(v-t)^2-r^2)n-1);
NF(1,K);
```

The answer 0 means that the condition (3.15) is also sufficient. We rediscovered the theorem:

Theorem 3.3. *Let $A = [0,0]$, $B = [a,0]$, $C = [x,y]$, $D = [u,v]$ be coordinates of four points A, B, C, D in a Cartesian coordinate system. A necessary and sufficient condition for a quadrilateral $ABCD$ to be cyclic is the condition*

$$-ayu + yu^2 + axv - x^2v - y^2v + yv^2 = 0. \qquad (3.16)$$

Remark 3.2. We can arrive at the condition (3.16) also by this consideration [67]. An arbitrary circle k has in any Cartesian coordinate system an equation $A_1(x^2 + y^2) + A_2x + A_3y + A_4 = 0$. Points $A = [x_1, y_1]$, $B = [x_2, y_2]$, $C = [x_3, y_3]$, $D = [x_4, y_4]$ lie on k if and only if they obey the system of equations $A_1(x_i^2 + y_i^2) + A_2x_i + A_3y_i + A_4 = 0$ for $i = 1, 2, 3, 4$. We have a system of four homogeneous equations with unknowns A_1, A_2, A_3, A_4 which has non-trivial solution iff the determinant of this system vanishes, i.e.,

$$\begin{vmatrix} x_1^2+y_1^2 & x_1 & y_1 & 1 \\ x_2^2+y_2^2 & x_2 & y_2 & 1 \\ x_3^2+y_3^2 & x_3 & y_3 & 1 \\ x_4^2+y_4^2 & x_4 & y_4 & 1 \end{vmatrix} = 0. \tag{3.17}$$

For the choice $A = [0,0]$, $B = [a,0]$, $C = [x,y]$, $D = [u,v]$ we get just the condition (3.16).

Now we will search for a necessary and sufficient condition for vertices of a quadrilateral $ABCD$ to lie on a circle, which is expressed in terms of distances between the vertices of $ABCD$.

Suppose that $ABCD$ is a quadrilateral with sides and diagonals of lengths $a = |AB|$, $b = |BC|$, $c = |CD|$, $d = |AD|$ and $e = |AC|, f = |BD|$ whose vertices lie on a circle centered at S and radius r. With the same notation as in the previous case (Fig. 3.5), it holds:

$r = |AS| \Leftrightarrow h_1 : s^2+t^2-r^2 = 0,$
$r = |BS| \Leftrightarrow h_2 : (a-s)^2+t^2-r^2 = 0,$
$r = |CS| \Leftrightarrow h_3 : (x-s)^2+(y-t)^2-r^2 = 0,$
$r = |DS| \Leftrightarrow h_4 : (u-s)^2+(v-t)^2-r^2 = 0,$
$b = |BC| \Leftrightarrow h_5 : (x-a)^2+y^2-b^2 = 0,$
$c = |CD| \Leftrightarrow h_6 : (u-x)^2+(v-y)^2-c^2 = 0,$
$d = |DA| \Leftrightarrow h_7 : u^2+v^2-d^2 = 0,$
$e = |AC| \Leftrightarrow h_8 : x^2+y^2-e^2 = 0,$
$f = |BD| \Leftrightarrow h_9 : (u-a)^2+v^2-f^2 = 0.$

In the ideal $I = (h_1, h_2, \dots, h_9)$ we eliminate all variables up to a, b, c, d, e to find out the length e in dependence on a, b, c, d. Analogously by elimination of all variables besides a, b, c, d, f we find the length f. For given side lengths a, b, c, d of a cyclic quadrilateral we get *two* lengths of diagonals e, f (Fig. 3.6). For a convex cyclic quadrilateral it follows

$$e^2(ab+cd) = (ac+bd)(ad+bc), \quad f^2(ad+bc) = (ac+bd)(ab+cd), \tag{3.18}$$

whereas for a non-convex quadrilateral we get

$$e^2(-ab+cd) = (ac-bd)(ad-bc), \quad f^2(-ad+bc) = (ac-bd)(ab-cd). \tag{3.19}$$

Mutual multiplication of the left and right sides in (3.18) gives in a convex case the relation

$$ef = ac+bd. \tag{3.20}$$

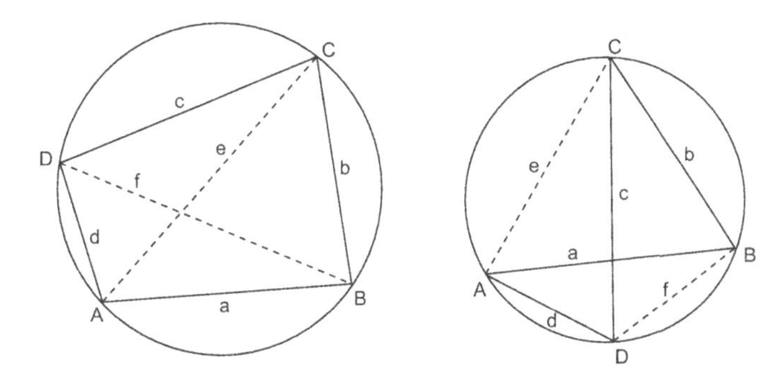

Fig. 3.6 Two cyclic quadrilaterals with the same side lengths a, b, c, d — convex and non-convex cases

Analogously in a non-convex case we get

$$ef = ac - bd \quad \text{or} \quad ef = -ac + bd. \tag{3.21}$$

We will call relations (3.20) and (3.21) *Ptolemy's conditions.*

From relations (3.18) and (3.19) we acquire another pair of interesting relations. Multiplication of the left sides by right sides and the right sides by left sides gives in a convex case[2]

$$e(ab + cd) = f(ad + bc) \tag{3.22}$$

and in a non-convex case

$$e(ab - cd) = f(ad - bc) \quad \text{or} \quad e(ab - cd) = f(-ad + bc). \tag{3.23}$$

It is easy to show that each of the Ptolemy's conditions $ef = ac + bd$, $ef = ac - bd$, $ef = -ac + bd$ is sufficient for a quadrilateral $ABCD$ to be cyclic. For instance for $ef = ac + bd$ we get

```
Use R::=Q[abcdefxyuvtz];
J:=Ideal((x-a)^2+y^2-b^2,(u-x)^2+(v-y)^2-c^2,u^2+v^2-d^2,x^2
+y^2-e^2,(u-a)^2+v^2-f^2,ac+bd-ef,az-1,(-yu^2+x^2v+y^2v-yv^2+
yua-xva)t-1);
NF(1,J);
```

the answer 0 which means that a quadrilateral $ABCD$ is cyclic.

Conversely, we will show that the Ptolemy's conditions $ef = ac + bd$, $ef = ac - bd$, $ef = -ac + bd$ are necessary for a quadrilateral $ABCD$ to be cyclic. We enter

[2](3.22) is surprisingly the necessary and sufficient condition for a convex quadrilateral with sides a, b, c, d and diagonals e, f to be cyclic, see [114, 100].

```
Use R::=Q[abcdefxyuvt];
K:=Ideal((x-a)^2+y^2-b^2,(u-x)^2+(v-y)^2-c^2,u^2+v^2-d^2,x^2
+y^2-e^2,(u-a)^2+v^2-f^2,-yu^2+x^2v+y^2v-yv^2+yua-xva,
(ac+bd-ef)(ac-bd+ef)(-ac+bd+ef)t-1);
NF(1,K);
```

and get the result 1 and the statement is not proved. Examination of non-degeneracy conditions does not give any additional conditions in the variables a, x, y, u, v. Whence, we eliminate independent variables x, y, u, v and a slack variable t in the ideal K. As a result we get three polynomials, one of which has the form $ac + bd + ef = 0$. Adding this condition to the *conclusion* of a statement (see the method B in Section 2.5) we get

```
Use R::=Q[abcdefxyuvt];
L:=Ideal((x-a)^2+y^2-b^2,(u-x)^2+(v-y)^2-c^2,u^2+v^2-d^2,x^2
+y^2-e^2,(u-a)^2+v^2-f^2,-yu^2+x^2v+y^2v-yv^2+yua-xva,
(ac-bd-ef)(ac+bd-ef)(ac-bd+ef)(ac+bd+ef)t-1);
NF(1,L);
```

the answer 0. We proved (and rediscovered) the well-known theorem of Ptolemy [67]:

Theorem 3.4 (Ptolemy). *A quadrilateral with side lengths a, b, c, d and diagonals e, f is cyclic if and only if*

$$(ac - bd - ef)(ac + bd - ef)(ac - bd + ef)(ac + bd + ef) = 0. \qquad (3.24)$$

Remark 3.3. The condition (3.24) can also be written in the form

$$\begin{vmatrix} 0 & a^2 & e^2 & d^2 \\ a^2 & 0 & b^2 & f^2 \\ e^2 & b^2 & 0 & c^2 \\ d^2 & f^2 & c^2 & 0 \end{vmatrix} = 0. \qquad (3.25)$$

Compare (3.25) with the formula (3.6). Namely it holds [5]:

Let $ABCD$ be a tetrahedron whose edges are of lengths $|AB| = a$, $|BC| = b$, $|CD| = c$, $|AD| = d$, $|AC| = e$, $|BD| = f$. Then for the radius r of the circumsphere of $ABCD$

$$\begin{vmatrix} 0 & a^2 & e^2 & d^2 \\ a^2 & 0 & b^2 & f^2 \\ e^2 & b^2 & 0 & c^2 \\ d^2 & f^2 & c^2 & 0 \end{vmatrix} = -2r^2 \begin{vmatrix} 0 & 1 & 1 & 1 & 1 \\ 1 & 0 & a^2 & e^2 & d^2 \\ 1 & a^2 & 0 & b^2 & f^2 \\ 1 & e^2 & b^2 & 0 & c^2 \\ 1 & d^2 & f^2 & c^2 & 0 \end{vmatrix} \qquad (3.26)$$

holds.

A similar formula holds for a triangle (here the relation analogous to (3.26) is equivalent to $p = abc/(4r)$ and in general for an arbitrary simplex. We will not give a proof of (3.26) — it could serve to readers as an exercise.

Let us show the derivation of relations (3.22) and (3.23) in a classical way, without the use of computer.

According to the well-known formula $p = abc/(4r)$ for the area a triangle with side lengths a, b, c and circumradius r for the area p of a cyclic quadrilateral $ABCD$ (Fig. 3.7),

$$p = abe/(4r) + cde/(4r)$$

holds, where we expressed the area p of a quadrilateral as the sum of areas of triangles ABC and ACD. If we express the area p as the sum of areas of triangles BCD and ABD then we get

$$p = bcf/(4r) + adf/(4r).$$

From the equality of the right sides the relation (3.22) follows. Analogously we prove (3.23).

3.2.2 *Formula of Heron*

Every triangle is cyclic since we can circumscribe a circle to any triangle. The area p of a triangle ABC with side lengths a, b, c is given by the formula of Heron

$$16p^2 = -a^4 - b^4 - c^4 + 2a^2b^2 + 2a^2c^2 + 2b^2c^2 \tag{3.27}$$

which we derived in Chapter 2 (relation (2.16)). By means of the formula (3.27) we will derive the relations below which we need later when investigating the area and radius of the circumcircle of a cyclic n-gon for $n > 3$.

We see that the right hand side of (3.27) is a symmetric polynomial in variables a^2, b^2, c^2, i.e., by any change of the order of variables a, b, c the formula (3.27) remains unchanged. Denote by k, l, m, the elementary symmetric functions in variables a^2, b^2, c^2, i.e.

$$k = a^2 + b^2 + c^2,\ l = a^2b^2 + b^2c^2 + c^2a^2,\ m = a^2b^2c^2. \tag{3.28}$$

Let $q = 16p^2, s = r^2$, where r is the circumradius of a triangle ABC. Then, according to the Fundamental Theorem of Symmetric Polynomials [24], the Heron formula (3.27) may be written in the form

$$k^2 - 4l + q = 0. \tag{3.29}$$

From the well-known formula $p = abc/(4r)$ for the area p of a triangle with side lengths a, b, c and the circumradius r we get

$$qs - m = 0. \tag{3.30}$$

This simple relation enables one to express q by means of s and vice versa.

To express the circumradius of a triangle in terms of its side lengths a, b, c, we substitute q from (3.29) into (3.30) which gives

$$s(k^2 - 4l) + m = 0. \tag{3.31}$$

3.2.3 *Formula of Brahmagupta*

As we mentioned in the previous section, we will use two basic methods to compute the area of a polygon. One of them is a coordinate method which requires an introduction of a coordinate system and uses the formula (3.3) for the area of a polygon.

The second method is a distance method which is based on the application of the Nagy–Rédey formula (3.4) and Ptolemy's conditions (3.20), (3.21). We will see that the use of the distance method is more effective. Let us solve the following problem:

Consider a cyclic quadrilateral $ABCD$ with side lengths a, b, c, d. Express the area p of a quadrilateral $ABCD$ in terms of a, b, c, d.

To find a formula for the area of $ABCD$ we will use both methods mentioned above — coordinate and distance. The use of the coordinate method is as follows.

Choose a Cartesian system of coordinates so that $A = [0, 0]$, $B = [a, 0]$, $C = [x, y]$, $D = [u, v]$ and the circumcircle of $ABCD$ is centered at the point $S = [s, t]$ with the radius r. With notation as in Fig. 3.7 we have:

$r = |AS| \Leftrightarrow h_1 : s^2 + t^2 - r^2 = 0,$
$r = |BS| \Leftrightarrow h_2 : (a - s)^2 + t^2 - r^2 = 0,$
$r = |CS| \Leftrightarrow h_3 : (x - s)^2 + (y - t)^2 - r^2 = 0,$
$r = |DS| \Leftrightarrow h_4 : (u - s)^2 + (v - t)^2 - r^2 = 0,$
$b = |BC| \Leftrightarrow h_5 : (x - a)^2 + y^2 - b^2 = 0,$
$c = |CD| \Leftrightarrow h_6 : (u - x)^2 + (v - y)^2 - c^2 = 0,$
$d = |DA| \Leftrightarrow h_7 : u^2 + v^2 - d^2 = 0.$

For the area p, we get, by (3.3),

$p =$ area of $ABCD \Leftrightarrow h_8 : p - 1/2(ay + xv - uy) = 0.$

Elimination of variables x, y, u, v, s, t, r in the ideal $I = (h_1, h_2, \ldots, h_8)$

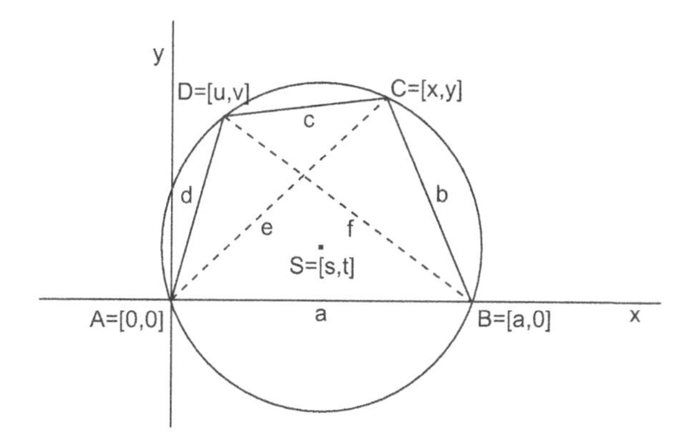

Fig. 3.7 Formula of Brahmagupta for the area of $ABCD$

```
Use R::=Q[xyuvstrabcdp];
I:=Ideal(s^2+t^2-r^2,(a-s)^2+t^2-r^2,(x-s)^2+(y-t)^2-r^2,
(u-s)^2+(v-t)^2-r^2,(x-a)^2+y^2-b^2,(u-x)^2+(v-y)^2-c^2,u^2
+v^2-d^2,p-1/2(ay+xv-yu));
Elim(x..r,I);
```

returns the only polynomial which leads after factorization to two equations. The first equation

$$16p^2 = -(a^4+b^4+c^4+d^4)+2(a^2b^2+a^2c^2+a^2d^2+b^2c^2+b^2d^2+c^2d^2)+8abcd, \quad (3.32)$$

which may be written as

$$16p^2 = (-a+b+c+d)(a-b+c+d)(a+b-c+d)(a+b+c-d), \quad (3.33)$$

gives the well-known formula of Brahmagupta for the area p of a *convex* cyclic quadrilateral.

The second equation

$$16p^2 = -(a^4+b^4+c^4+d^4)+2(a^2b^2+a^2c^2+a^2d^2+b^2c^2+b^2d^2+c^2d^2)-8abcd \quad (3.34)$$

expresses the area of a *non-convex* cyclic quadrilateral.

The relation (3.34) can be written in the form

$$16p^2 = (-a+b-c+d)(a-b-c+d)(a+b+c+d)(a+b-c-d) \quad (3.35)$$

which is analogous to (3.33).

Remark 3.4.

1) We can arrive at the formula (3.34) from (3.32), if we write $-c$ in (3.32)

instead of c.

2) We should realize that all the derived formulas are *necessary* conditions for the area p. For instance, for a choice $a = 2, b = 1, c = 1, d = 1$ we get only one (real) solution since a non-convex quadrilateral obviously does not exist.

Elimination of variables x, y, u, v, s, t in the ideal $J = (h_1, h_2, \dots, h_7)$

```
Use R::=Q[xyuvstrabcd];
J:=Ideal(s^2+t^2-r^2,(a-s)^2+t^2-r^2,(x-s)^2+(y-t)^2-r^2,
(u-s)^2+(v-t)^2-r^2,(x-a)^2+y^2-b^2,(u-x)^2+(v-y)^2-c^2,u^2
+v^2-d^2);
Elim(x..t,J);
```

gives the radius r of the circumcircle of a cyclic quadrilateral with side lengths a, b, c, d.

For a convex cyclic quadrilateral we get

$$r^2 = \frac{(ab+cd)(ac+bd)(ad+bc)}{(a+b+c-d)(a+b-c+d)(a-b+c+d)(-a+b+c+d)} \tag{3.36}$$

and for a non-convex cyclic quadrilateral

$$r^2 = \frac{(ab-cd)(-ac+bd)(ad-bc)}{(-a+b-c+d)(a-b-c+d)(a+b+c+d)(a+b-c-d)}. \tag{3.37}$$

Hence, for given side lengths a, b, c, d, (which obey the analogue of the triangle inequality $a + b + c > d, \dots,$ etc.), there exist *at most* two cyclic quadrilaterals with *different* areas and circumradii (Fig. 3.6).

The second and more effective way to explore the area of a cyclic quadrilateral is the distance method. In order to express the area p of a quadrilateral $ABCD$ we will use the formula of Staudt (3.5).

Considering cyclic quadrilaterals, then by the Ptolemy's condition for convex quadrilaterals (3.20) the relation $ef = ac + bd$ holds. Using the formula of Staudt (3.5) followed by the elimination of variables e, f gives

```
Use R::=Q[abcdefp];
I:=Ideal(16p^2-4e^2f^2+(a^2-b^2+c^2-d^2)^2,ac+bd-ef);
Elim(e..f,I);
```

the Brahmagupta's relation (3.32) for convex cyclic quadrilaterals.

Similarly, from the Ptolemy's conditions (3.21) $ef = ac - bd,\ ef = bd - ac$ for non-convex quadrilaterals and the formula of Staudt we get

```
Use R::=Q[abcdefp];
J:=Ideal(16p^2-4e^2f^2+(a^2-b^2+c^2-d^2)^2,
(ac-bd-ef)(ac-bd+ef));
Elim(e..f,J);
```

the Brahmagupta's relation (3.34) for non-convex cyclic quadrilaterals.

We see that both formulas (3.32), (3.34) are symmetric in variables a^2, b^2, c^2, d^2. Replacing (3.32), (3.34) by the elementary symmetric functions k, l, m, n, where

$$\begin{aligned} k &= a^2 + b^2 + c^2 + d^2, \\ l &= a^2b^2 + a^2c^2 + a^2d^2 + b^2c^2 + b^2d^2 + c^2d^2, \\ m &= a^2b^2c^2 + a^2b^2d^2 + a^2c^2d^2 + b^2c^2d^2, \\ n &= a^2b^2c^2d^2, \end{aligned} \tag{3.38}$$

we get the following important formula (3.39), which involves both relations (3.32), (3.34)

$$(k^2 - 4l + q)^2 - 64n = 0, \tag{3.39}$$

where we denoted $q = 16p^2$.

Remark 3.5.
1) Note that if some of the side lengths a, b, c, d vanishes, then $n = 0$ and (3.39) becomes the formula of Heron (3.29).

2) Setting

```
Use R::=Q[abcdefp];
K:=Ideal(16p^2-4e^2f^2+(a^2-b^2+c^2-d^2)^2,
(ac+bd-ef)(ac-bd-ef)(ac-bd+ef));
Elim(e..f,K);
```

we obtain both Brahmagupta's formulas (3.32), (3.34) all at once as a product of two polynomials.

To compute the circumradius r we can use, unlike the coordinate method, the following procedure [112].

We divide a cyclic quadrilateral $ABCD$ by a diagonal $e = |AC|$ into two triangles ABC and CDA with the sides a, b, e and c, d, e respectively

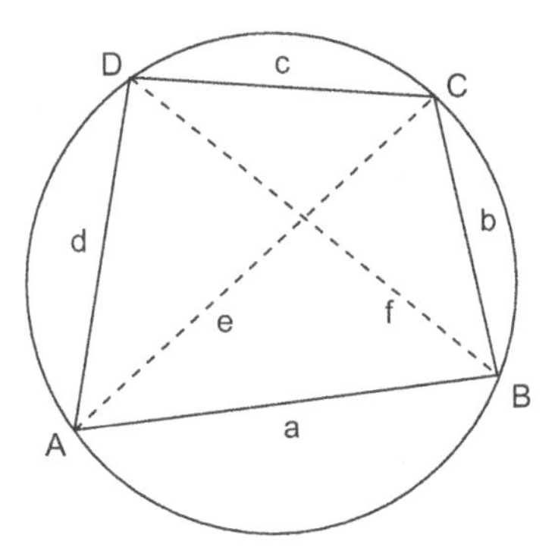

Fig. 3.8 Computing the circumradius of a cyclic quadrilateral $ABCD$ with sides a, b, c, d — a convex case

(Fig. 3.8). Both triangles have the same circumradius r as a quadrilateral $ABCD$. By (3.31) for the circumradius of a triangle we eliminate the common length e in the triangles ABC, CDA

```
Use R::=Q[abcdeklmns];
I:=Ideal(s((a^2+b^2+e^2)^2-4(a^2b^2+b^2e^2+e^2a^2))+a^2b^2e^2,
s((c^2+d^2+e^2)^2-4(c^2d^2+d^2e^2+e^2c^2))+c^2d^2e^2,a^2+b^2+
c^2+d^2-k,a^2b^2+a^2c^2+a^2d^2+b^2c^2+b^2d^2+c^2d^2-l,
a^2b^2c^2+a^2b^2d^2+a^2c^2d^2+b^2c^2d^2-m,a^2b^2c^2d^2-n);
Elim(a..e,I);
```

which leads to the formula for computing the circumradius r of a cyclic quadrilateral $ABCD$

$$(s(k^2 - 4l) + m)^2 - n(8s - k)^2 = 0, \tag{3.40}$$

where k, l, m, n are the elementary symmetric functions (3.38) and $s = r^2$. This formula involves both relations (3.36), (3.37) for computation of the circumradius of a cyclic quadrilateral.

The following formula gives a relation between s and q

$$(qs - m)^2 - k^2 n = 0. \tag{3.41}$$

Note again that if $n = 0$, then a quadrilateral transforms into a triangle and (3.40), (3.41) become the formulas (3.31) and (3.30).

Now we show classical proofs of formulas (3.33) and (3.35).

First we consider a convex case. Suppose that a convex cyclic quadrilateral $ABCD$ with side lengths a, b, c, d is given (Fig. 3.9). From the

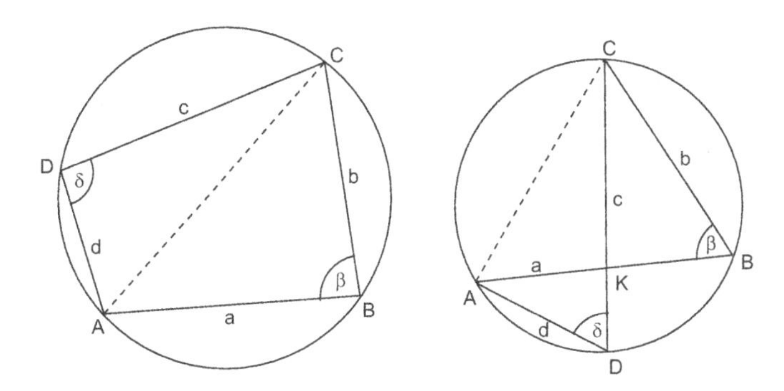

Fig. 3.9 Cyclic quadrilaterals with sides a, b, c, d — convex and non-convex cases

triangles ABC and ADC we express the length of a diagonal $|AC|$ by the law of cosines as

$$|AC|^2 = a^2 + b^2 - 2ab\cos\beta$$

and

$$|AC|^2 = c^2 + d^2 - 2cd\cos\delta.$$

The equality of the right sides implies

$$a^2 + b^2 - c^2 - d^2 = 2ab\cos\beta - 2cd\cos\delta,$$

and from here, with respect to the fact that in a cyclic quadrilateral $\beta + \delta = 180°$, we get

$$a^2 + b^2 - c^2 - d^2 = 2(ab + cd)\cos\beta. \tag{3.42}$$

Furthermore, the area p of a quadrilateral $ABCD$ is the sum of the areas of triangles ABC and ADC, i.e., $p = 1/2ab\sin\beta + 1/2cd\sin\delta$.
Since $\beta + \delta = 180°$, then $\sin\beta = \sin\delta$ and

$$p = \frac{1}{2}(ab + cd)\sin\beta. \tag{3.43}$$

We raise the equalities (3.42) and (3.43) to the second power, and then the substitution of (3.42) into (3.43), with the use of the relation $\sin^2\beta = 1 - \cos^2\beta$, gives

$$\begin{aligned}
16p^2 &= 4(ab + cd)^2 - (a^2 + b^2 - c^2 - d^2)^2 \\
&= (2(ab + cd) + (a^2 + b^2 - c^2 - d^2))(2(ab + cd) - (a^2 + b^2 - c^2 - d^2)) \\
&= ((a + b)^2 - (c - d)^2)(-(a - b)^2 + (c + d)^2) \\
&= (a + b + c - d)(a + b - c + d)(a - b + c + d)(-a + b + c + d),
\end{aligned}$$

which is (3.33).

Now we will prove the formula (3.35) for the area of a non-convex cyclic quadrilateral knowing the side lengths a, b, c, d. For the same side lengths a, b, c, d a non-convex cyclic quadrilateral $ABCD$ is constructed in Fig. 3.9. Similarly as in the convex case we express the length of the diagonal AC by the law of cosines in two ways. Taking into account that the angles β and δ are equal, we get

$$a^2 + b^2 - c^2 - d^2 = 2(ab - cd)\cos\beta. \tag{3.44}$$

The signed area p of a quadrilateral $ABCD$ equals the difference of the areas of triangles BCK and DAK which is equal to the difference of the areas of triangles ABC and CDA.

Notice, that when we move along the perimeter of ABC in the order of vertices from A through B to C, we proceed anti-clockwise, i.e., in the positive sense, whereas going along the perimeter of CDA, we move in the negative sense. That is why the area of a triangle ABC is considered with the sign plus and the area of a triangle CDA with the sign minus. We get

$$p = \frac{1}{2}(ab - cd)\sin\beta. \tag{3.45}$$

We raise the equalities (3.44) and (3.45) to the second power and by the substitution (3.44) into (3.45), similarly as in the previous case, we get

$$\begin{aligned}
&16p^2 = 4(ab - cd)^2 - (a^2 + b^2 - c^2 - d^2)^2 \\
&= (2(ab - cd) + (a^2 + b^2 - c^2 - d^2))(2(ab - cd) - (a^2 + b^2 - c^2 - d^2)) \\
&= ((a + b)^2 - (c + d)^2)(-(a - b)^2 + (c - d)^2) \\
&= (a + b + c + d)(a + b - c - d)(a - b + c - d)(-a + b + c - d),
\end{aligned}$$

which is the formula (3.35).

3.2.4 *Area of a cyclic pentagon*

Now we are ready to investigate the area p of a cyclic pentagon with the side lengths a, b, c, d, e. We will proceed in a similar way as with the previous sections. First we will use the coordinate method to compute the area of a cyclic pentagon.

Choose a Cartesian coordinate system so that for the vertices of a cyclic pentagon $A = [0, 0]$, $B = [a, 0]$, $C = [x, y]$, $D = [u, v]$, $E = [w, z]$. Let $S = [s, t]$ be coordinates of the circumcenter S and r the circumradius. Further denote $a = |AB|$, $b = |BC|$, $c = |CD|$, $d = |DE|$, $e = |EA|$ (Fig. 3.10). Then

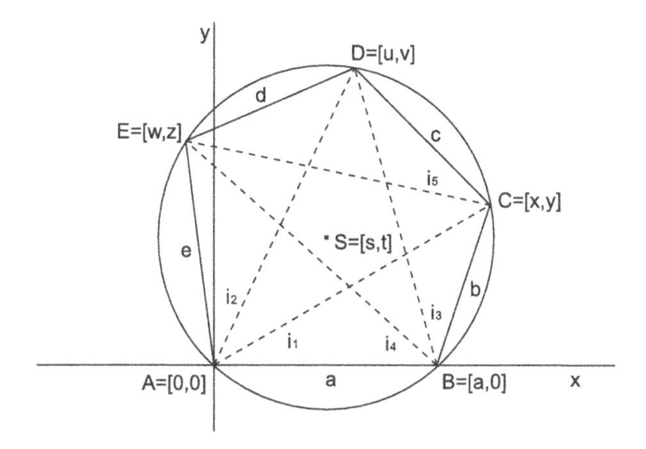

Fig. 3.10 Area of a cyclic pentagon — convex case

$$\begin{aligned}
b = |BC| &\Leftrightarrow h_1 : (x-a)^2 + y^2 - b^2 = 0,\\
c = |CD| &\Leftrightarrow h_2 : (u-x)^2 + (v-y)^2 - c^2 = 0,\\
d = |DE| &\Leftrightarrow h_3 : (w-u)^2 + (z-v)^2 - d^2 = 0,\\
e = |EA| &\Leftrightarrow h_4 : w^2 + z^2 - e^2 = 0,\\
r = |SA| &\Leftrightarrow h_5 : s^2 + t^2 - r^2 = 0,\\
r = |SB| &\Leftrightarrow h_6 : (a-s)^2 + t^2 - r^2 = 0,\\
r = |SC| &\Leftrightarrow h_7 : (x-s)^2 + (y-t)^2 - r^2 = 0,\\
r = |SD| &\Leftrightarrow h_8 : (u-s)^2 + (v-t)^2 - r^2 = 0,\\
r = |SE| &\Leftrightarrow h_9 : (w-s)^2 + (z-t)^2 - r^2 = 0.
\end{aligned}$$

By (3.3) for the area p of $ABCDE$

$$p = \text{area of } ABCDE \Leftrightarrow h_{10} : p - 1/2(ay + xv - yu + uz - vw) = 0.$$

We have the ideal $I = (h_1, h_2, \ldots, h_{10})$ given with 10 polynomials. To express the area p of $ABCDE$ in terms of its side lengths a, b, c, d, e we eliminate 9 variables $x, y, u, v, w, z, s, t, r$ in I.

```
Use R::=Q[xyuvwzstrabcdep];
I:=Ideal((x-a)^2+y^2-b^2,(u-x)^2+(v-y)^2-c^2,(w-u)^2+(z-v)^2-
d^2,w^2+z^2-e^2,s^2+t^2-r^2,(a-s)^2+t^2-r^2,(x-s)^2+(y-t)^2-
r^2,(u-s)^2+(v-t)^2-r^2,(w-s)^2+(z-t)^2-r^2,ay+xv-yu+uz-vw-2p);
Elim(x..r,I);
```

In 9 hours and 5 minutes we get in CoCoA the only equation in the form of a polynomial of 14th degree in p which contains 6672 terms.

Substitution of the elementary symmetric functions

$$\begin{aligned} k &= \sum a^2 = a^2 + b^2 + \cdots + e^2, \\ l &= \sum a^2 b^2 = a^2 b^2 + a^2 c^2 + \cdots + d^2 e^2, \\ m &= \sum a^2 b^2 c^2 = a^2 b^2 c^2 + a^2 b^2 d^2 + \cdots + c^2 d^2 e^2, \\ n &= \sum a^2 b^2 c^2 d^2 = a^2 b^2 c^2 d^2 + a^2 b^2 c^2 e^2 + \cdots + b^2 c^2 d^2 e^2, \\ o &= a^2 b^2 c^2 d^2 e^2, \end{aligned} \tag{3.46}$$

with $q = 16p^2$ leads to the equation $h = 0$ which contains 153 terms and is still too long to write it completely. It starts with

$$h : q^7 + q^6(7k^2 - 24l) + q^5(21k^4 - 144k^2 l + 240l^2 + 16km - 192n) + \cdots = 0.$$

Now we introduce new variables g, L, M, N given by

$$h_1 : k^2 - 4l + q - g = 0,$$

$$h_2 : kg + 8m - L = 0,$$

$$h_3 : g^2 - 64n - M = 0,$$

$$h_4 : 128o - N = 0.$$

Then the elimination of variables k, l, m, n, o, g in the ideal (h, h_1, h_2, h_3, h_4) leads to the formula (3.47) which contains only 5 terms. We arrived at the following theorem [112]:

Theorem 3.5 (Robbins). *Given a cyclic pentagon with the side lengths a, b, c, d, e and the area p. Let $q = 16p^2$ and L, M, N be as above. Then q obeys the equation*

$$L^2 M^2 + M^3 q - 16L^3 N - 18LMNq - 27N^2 q^2 = 0. \tag{3.47}$$

The relation (3.47) can be considered as a generalization of the formulas of Heron and Brahmagupta.

Now we derive the formula (3.47) by the distance method. We use the Nagy–Rédey formula (3.4) for the area of a pentagon which is given by its side lengths a, b, c, d, e and diagonals i_1, i_2, i_3, i_4, i_5. In the situation in Fig. 3.10, first we apply the Ptolemy's theorem to cyclic quadrilaterals $ABCD$, $BCDE$, $CDEA$, $DEAB$, $EABC$ and get

$$\begin{aligned} &i_1 i_3 = ac + bi_2, \;\; i_3 i_5 = bd + ci_4, \;\; i_5 i_2 = ce + di_1, \;\; i_2 i_4 = da + ei_3, \\ &i_4 i_1 = eb + ai_5. \end{aligned} \tag{3.48}$$

Substitution of conditions (3.48) into the Nagy–Rédey formula (3.13) for a pentagon gives

$$16p^2 = -\sum a^4 + 2\sum a^2b^2 + g, \tag{3.49}$$

where

$$g = 4(abci_2 + bcdi_4 + cdei_1 + deai_3 + eabi_5). \tag{3.50}$$

If we replace the formula (3.49) by the elementary symmetric functions k, l from (3.46) and $q = 16p^2$, then (3.49) can be written in the form

$$k^2 - 4l + q = g. \tag{3.51}$$

Hence, in order to express the area p of a cyclic pentagon $ABCDE$ in terms of a, b, c, d, e, it suffices to express g from (3.50) by means of a, b, c, d, e, because other parts *do not depend* on variables i_1, i_2, i_3, i_4, i_5.
The elimination of i_1, i_2, i_3, i_4, i_5 from g requires, besides Ptolemy's conditions (3.48), other relations which hold between sides and diagonals of a cyclic pentagon. Namely, the conditions (3.48) are mutually dependent, while only three of them are independent. Relations which we need are similar to those in (3.22) and in our case they have the form

$$\begin{aligned} &i_1(ab + ci_2) = i_3(bc + ai_2),\ i_3(bc + di_4) = i_5(cd + bi_4),\\ &i_5(cd + ei_1) = i_2(ed + ci_1),\ i_2(ed + ai_3) = i_4(ea + di_3),\\ &i_4(ea + bi_5) = i_1(ab + ei_5). \end{aligned} \tag{3.52}$$

We call the conditions (3.52) *additional Ptolemy's conditions.* Elimination of variables i_1, i_2, i_3, i_4, i_5 in the ideal I generated by a collection of polynomials (3.48), (3.52) and (3.50) is now feasible.

```
Use R::=Q[abcdei[1..5]g];
I:=Ideal(ac+bi[2]-i[1]i[3],bd+ci[4]-i[3]i[5],ce+di[1]-i[5]i[2],
da+ei[3]-i[2]i[4],eb+ai[5]-i[4]i[1],i[1](ab+ci[2])-i[3](bc+
ai[2]),i[3](bc+di[4])-i[5](cd+bi[4]),i[5](cd+ei[1])-i[2](ed+
ci[1]),4(abci[2]+bcdi[4]+cdei[1]+deai[3]+eabi[5])-g);
Elim(i[1]..i[5],I);
```

In 9 minutes and 18 seconds in CoCoA we get the polynomial $F_1(g, a, b, c, d, e)$ which contains 827 terms. Substitution of the elementary symmetric functions k, l, m, n, o from (3.46) into F_1 gives the polynomial $F_2(g, k, l, m, n, o)$ with 37 terms.

Another substitution

$$g = k^2 - 4l + q,\ L = kg + 8m,\ M = g^2 - 64n,\ N = 128o$$

into F_2, which can be carried out even by hand, yields the Robbins' formula (3.47).

Until now we explored the area of a *convex* cyclic pentagon. Now let us consider a non-convex cyclic pentagon with the same side lengths a, b, c, d, e (see Fig. 3.11). Then the Ptolemy's conditions are

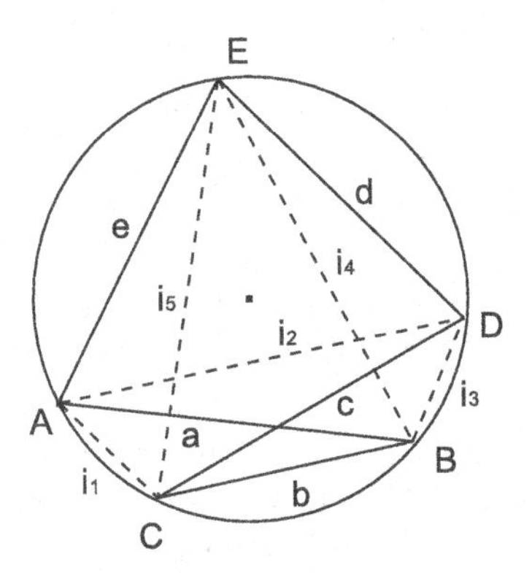

Fig. 3.11 Area of a cyclic pentagon — non-convex case

$$\begin{aligned} &i_1 i_3 = ac - bi_2, \; i_3 i_5 = -bd + ci_4, \; i_5 i_2 = ce + di_1, \; i_2 i_4 = da + ei_3, \\ &i_4 i_1 = -eb + ai_5. \end{aligned} \tag{3.53}$$

By (3.23) we get the following additional Ptolemy's conditions:

$$\begin{aligned} &i_1(ab - ci_2) = i_3(bc - ai_2), \; i_3(bc - di_4) = i_5(-cd + bi_4), \\ &i_5(cd + ei_1) = i_2(de + ci_1), \; i_2(de + ai_3) = i_4(ea + di_3), \\ &i_4(ea - bi_5) = i_1(-ab + ei_5). \end{aligned} \tag{3.54}$$

Instead of (3.50) we get

$$g = -abci_2 - bcdi_4 + cdei_1 + deai_3 - eabi_5. \tag{3.55}$$

Writing $-b$ instead of b in (3.53), (3.54) and (3.55), we get the same system of equations as in the convex case. Thus, the elimination of variables i_1, i_2, i_3, i_4, i_5 in (3.55) gives the same result. Similarly we proceed in other cases.

Now we will summarize the main steps of the previous procedure:

Cyclic Pentagon Area Algorithm:

1. Eliminate the lengths of diagonals i_1, i_2, i_3, i_4, i_5 from the system (3.50),

(3.48) and (3.52) to obtain the symmetric polynomial $F_1(g, a, b, c, d, e)$.

2. Express the polynomial $F_1(g, a, b, c, d, e)$ by the elementary symmetric functions k, l, m, n, o to get the polynomial $F_2(g, k, l, m, n, o)$.

3. Eliminate g, k, l, m, n, o in the ideal (F_2, L, M, N) to obtain the final polynomial $F_3(q, L, M, N)$.

Remark 3.6.
1) The polynomial (3.47) is of 7th degree in $q = 16p^2$. This means that there exist *at most* 7 cyclic pentagons with the given side lengths a, b, c, d, e.

2) The terms g, L, M in (3.47) have a geometric meaning. g is the left side of (3.29), L fulfils the equation $8qs - L = 0$ which is equivalent to (3.41), and M is the left side of the formula (3.39).

3) The formula (3.47) was published by D. P. Robbins in 1994 (see [112]). His discovery is based on the fact that the left-hand side of (3.47) is the discriminant of the cubic equation $x^3 + 2Lx^2 - Mqx + 2Nq^2 = 0$. Why it is the case is still a mystery.

3.2.5 *Radius of a cyclic pentagon*

Now we will compute the circumradius r of a cyclic pentagon $ABCDE$ with the given side lengths a, b, c, d, e.

Considering the coordinate method, we introduce a Cartesian coordinate system so that $A = [r, 0]$, $B = [x, y]$, $C = [u, v]$, $D = [w, z]$, $E = [s, t]$. Place the origin O into the circumcenter (Fig. 3.12). Eliminating 8 variables x, y, u, v, w, z, s, t in the respective ideal, we get in 11 minutes and 5 seconds

```
Use R::=Q[xyuvwzstabcder];
I:=Ideal((x-r)^2+y^2-a^2,(u-x)^2+(v-y)^2-b^2,(w-u)^2+(z-v)^2-
c^2,(s-w)^2+(t-z)^2-d^2,(s-r)^2+t^2-e^2,x^2+y^2-r^2,u^2+v^2-
r^2,w^2+z^2-r^2,s^2+t^2-r^2);
Elim(x..t,I);
```

a polynomial $G_1(s, a, b, c, d, e)$ of 7th degree in $s = r^2$ with 2992 terms. Substitution of the elementary symmetric functions k, l, m, n, o from (3.46) yields the polynomial $G_2(s, k, l, m, n, o)$ with 81 terms in (3.56).

The distance method is as follows. Divide a cyclic pentagon $ABCDE$ into three triangles ABC, ACD and ADE (see Fig. 3.12). It is clear that all

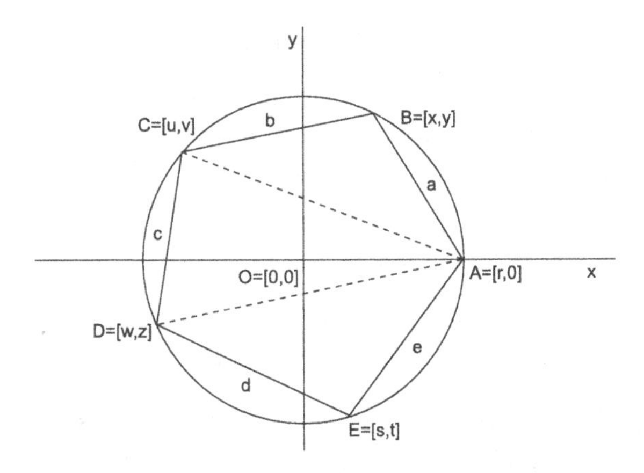

Fig. 3.12 Radius of a cyclic pentagon

these triangles have the same circumradius. We express the circumradius r in each of the triangles ABC, ACD, ADE by (3.31), and then we eliminate the common diagonals i_1, i_2. We enter

```
Use R::=Q[abcdei[1..2]s];
J:=Ideal(s((a^2+b^2+i[1]^2)^2-4(a^2b^2+a^2i[1]^2+b^2i[1]^2))
+a^2b^2i[1]^2,s((c^2+i[1]^2+i[2]^2)^2-4(c^2i[1]^2+c^2i[2]^2+
i[1]^2i[2]^2))+c^2i[1]^2i[2]^2,s((d^2+e^2+i[2]^2)^2-4(d^2e^2
+d^2i[2]^2+e^2i[2]^2))+d^2e^2i[2]^2);
Elim(i[1]..i[2],J);
```

and in 17 seconds in CoCoA we obtain (3.56). We can state the following theorem:

Theorem 3.6. *A pentagon with the sides a, b, c, d, e which is inscribed in the circle with the radius r is given. Let s, k, l, m, n, o be as above. Then*

$$s^3[(s(k^2-4l)+m)^2-n(8s-k)^2]^2+os^2Q+o^2sP+o^3=0 \qquad (3.56)$$

holds, where P, Q are polynomials in k, l, m, n.

The main steps to find the circumradius of a cyclic pentagon are as follows:

Cyclic Pentagon Radius Algorithm:

1. Divide a cyclic pentagon $ABCDE$ into three triangles ABC, ACD, ADE and write the respective formulas (3.31) for ABC, ACD, ADE.

2. Eliminate i_1, i_2 from the system of equations from the first step to obtain

the polynomial $G_1(s, a, b, c, d, e)$.

3. Eliminate a, b, c, d, e from the system of G_1 and the elementary symmetric functions k, l, m, n, o from (3.46) to obtain the polynomial $G_2(s, k, l, m, n, o)$.

Remark 3.7.
1) There exist at most 7 cyclic pentagons with different radii.

2) If we put $o = 0$ in (3.56) we get a quadrilateral and the formula (3.56) becomes (3.40).

3.2.6 *Area of a cyclic hexagon*

In this section we try to apply the previous distance method to express the area of a cyclic hexagon in terms of its side lengths.

Consider a cyclic hexagon $ABCDEF$ with side lengths $a = |AB|$, $b = |BC|$, $c = |CD|$, $d = |DE|$, $e = |EF|$, $f = |FA|$ and diagonals $i_1 = |AC|$, $i_2 = |AD|$, $i_3 = |AE|$, $i_4 = |BD|$, $i_5 = |BE|$, $i_6 = |BF|$, $i_7 = |CE|$, $i_8 = |CF|$, $i_9 = |DF|$ (Fig. 3.13).

The following Ptolemy's conditions (3.57) hold:

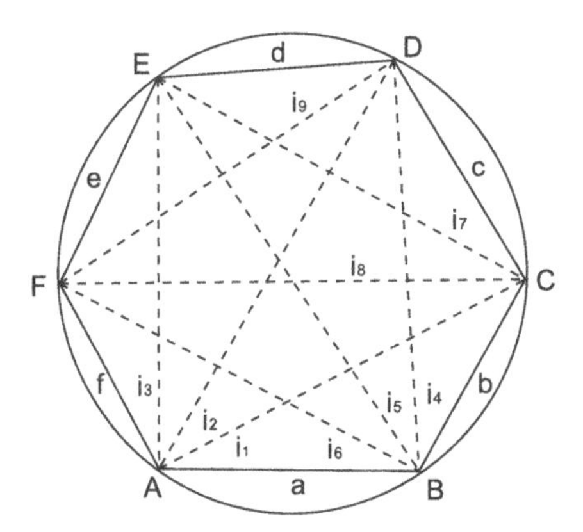

Fig. 3.13 Area of a cyclic hexagon

$$\begin{aligned}
&i_1 i_4 = ac + bi_2,\ i_4 i_7 = bd + ci_5,\ i_7 i_9 = ce + di_8,\ i_9 i_3 = df + ei_2,\\
&i_3 i_6 = ea + fi_5,\ i_6 i_1 = fb + ai_8,\ i_2 i_5 = ad + i_3 i_4,\ i_5 i_8 = be + i_6 i_7,\\
&i_2 i_8 = cf + i_1 i_9. \qquad (3.57)
\end{aligned}$$

From the Nagy–Rédey formula (3.14) for the area p of a hexagon $ABCDEF$ together with the use of the Ptolemy's conditions (3.57), we obtain, similarly as in the case of a pentagon, the relation

$$16p^2 = -\sum a^4 + 2\sum a^2b^2 + g, \tag{3.58}$$

which has the same form as (3.49), and which may be briefly written in the form

$$k^2 - 4l + q = g, \tag{3.59}$$

where

$$g = 4(abci_2+bcdi_5+cdei_8+defi_2+efai_5+fabi_8+adi_3i_4+bei_6i_7+cfi_1i_9),$$

$q = 16p^2$ and

$$\begin{aligned}
k &= \sum a^2 = a^2 + b^2 + \cdots + f^2,\\
l &= \sum a^2b^2 = a^2b^2 + a^2c^2 + \cdots + e^2f^2,\\
m &= \sum a^2b^2c^2 = a^2b^2c^2 + a^2b^2d^2 + \cdots + d^2e^2f^2,\\
n &= \sum a^2b^2c^2d^2 = a^2b^2c^2d^2 + a^2b^2c^2e^2 + \cdots + c^2d^2e^2f^2,\\
o &= \sum a^2b^2c^2d^2e^2 = a^2b^2c^2d^2e^2 + a^2b^2c^2d^2f^2 + \cdots + b^2c^2d^2e^2f^2,\\
t &= a^2b^2c^2d^2e^2f^2
\end{aligned} \tag{3.60}$$

are elementary symmetric functions in $a^2, b^2, c^2, d^2, e^2, f^2$.

To get a formula for the area of a cyclic hexagon in terms of its side lengths a, b, c, d, e, f, we eliminate variables $i_1, i_2, \ldots, i_9$ in g, since all other expressions in (3.59) contain only variables a, b, c, d, e, f. This elimination requires adding to the respective ideal, besides Ptolemy's conditions (3.57), the following additional Ptolemy's conditions (3.61) which have a similar form as (3.52):

$$\begin{aligned}
&i_1(ab + ci_2) = i_4(bc + ai_2),\ i_4(bc + di_5) = i_7(cd + bi_5),\\
&i_7(cd + ei_8) = i_9(de + ci_8),\ i_9(de + fi_2) = i_3(ef + di_2),\\
&i_3(ef + ai_5) = i_6(fa + ei_5),\ i_6(fa + bi_8) = i_1(ab + fi_8),\\
&i_2(ai_4 + di_3) = i_5(di_4 + ai_3),\ i_5(bi_7 + ei_6) = i_8(bi_6 + ei_7),\\
&i_8(ci_9 + fi_1) = i_2(ci_1 + fi_9).
\end{aligned} \tag{3.61}$$

Here our investigation ends. Because of a big complexity it was not possible to carry out indicated elimination. It is necessary to look for another simplification.

We obtain similar results by computing the circumradius of a cyclic hexagon.

3.3 Final remarks

1. Two given algorithms Cyclic Pentagon Area Algorithm and Cyclic Pentagon Radius Algorithm could serve as a tool for computing the area and radius of a cyclic n-gon for $n \geq 6$.

2. Using this method two main problems occurred:

a) A huge amount of CPU time is spent for computations.
b) Finding appropriate expressions like L, M, N in the Robbins' formula (3.47) to abbreviate the final polynomial proved to be less easy than it seems.

3. Consider an arbitrary cyclic $(2n+1)$-gon with the given sides $a_1, a_2, \ldots, a_{2n+1}$. Then it was conjectured by Robbins [112] that there exist at most k_n such $(2n+1)$-gons with different areas and radii, where

$$k_n = \sum_{j=0}^{n-1} (n-j) \binom{2n+1}{j}, \tag{3.62}$$

i.e., $k_1, k_2, k_3, \cdots = 1, 7, 38, 187, 874, \ldots$

For $(2n+2)$-gons with sides $a_1, a_2, \ldots, a_{2n+2}$ there are at most $2k_n$ such polygons with different areas and radii.

This conjecture was confirmed in 2004 (see [38, 71]). In [71], corresponding formulas for inscribed heptagon and octagon are given as well.

4. The formula which connects the area and radius of a cyclic pentagon of the type (3.30), (3.41) is still missing.

Chapter 4

Simson–Wallace theorem

There is a nice property of the circumcircle of a triangle, which is often ascribed to the Scottish mathematician R. Simson (1687–1768), but it was really discovered by another Scottish mathematician W. Wallace (1768–1843) in 1799 (see [26]). Therefore it is quite natural to call the following statement the Simson–Wallace theorem (Fig. 4.1):

Theorem 4.1 (Simson–Wallace). *Let ABC be a triangle and P a point of the circumcircle of ABC. Then the feet of perpendiculars from P onto the sides of ABC lie on a straight line.*

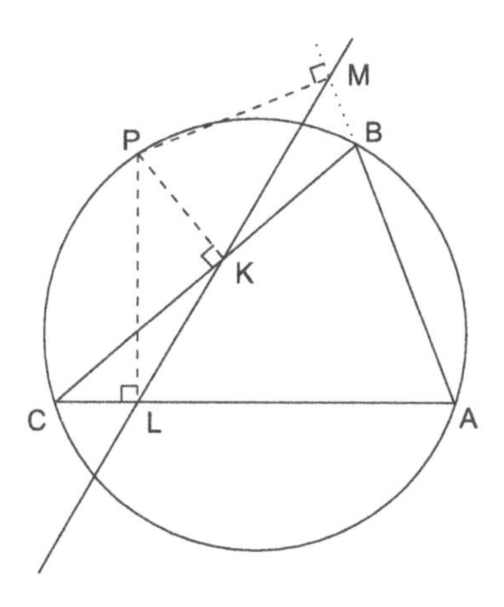

Fig. 4.1 Points K, L, M are collinear

The theorem shows an interesting property of points which lie on the circumcircle of a triangle. If a point P moves along the circumcircle of ABC, then the respective feet of perpendiculars to the sides of a triangle will also move to form different straight lines. Thus to every point P of the circumcircle we can assign by the Simson–Wallace theorem the straight line which we call Simson–Wallace line. This line has many interesting properties, see for example [26, 72, 124].

By dynamic software Cabri II Plus we may *verify* the theorem asking a question whether the feet K, L, M of perpendiculars from a point P to the sides of a triangle are collinear. We obtain the answer "yes". However, this response is not based on a logical consideration but on the fact that the construction in Cabri II Plus is very precise. Remember that this verification *does not replace* a proof.

First we will prove the Simson–Wallace theorem by automatic theorem proving.

We express the hypotheses and the conclusion analytically. Choose a Cartesian system of coordinates so that $A = [a, 0]$, $B = [b, c]$, $C = [0, 0]$, $P = [p, q]$, $K = [k_1, k_2]$, $L = [l_1, 0]$, $M = [m_1, m_2]$ (Fig. 4.2).
The hypotheses are as follows:

$PL \perp AC \Leftrightarrow h_1 : p - l_1 = 0,$

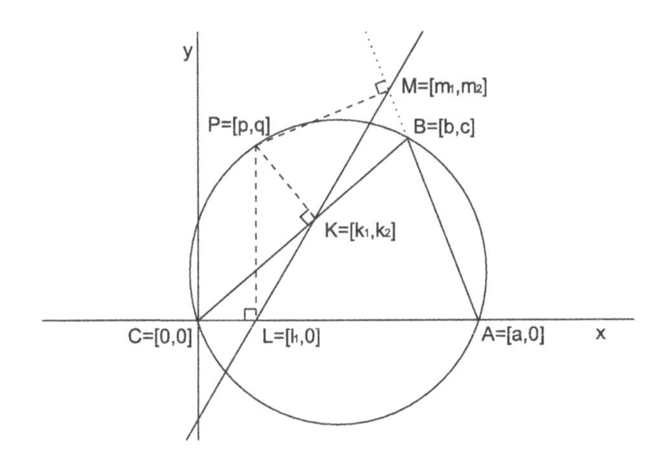

Fig. 4.2 Simson–Wallace theorem — computer proof

$K \in BC \Leftrightarrow h_2 : ck_1 - bk_2 = 0,$
$PK \perp BC \Leftrightarrow h_3 : (p - k_1)b + (q - k_2)c = 0,$
$M \in AB \Leftrightarrow h_4 : ac + bm_2 - cm_1 - am_2 = 0,$
$PM \perp AB \Leftrightarrow h_5 : (p - m_1)(b - a) + (q - m_2)c = 0,$

By (3.15) from Chapter 3 we get the condition

P lies on the circumcircle of $ABC \Leftrightarrow$

$h_6 : -acp + cp^2 + abq - b^2q - c^2q + cq^2 = 0.$

The conclusion z has the form:

K, L, M are collinear $\Leftrightarrow z : l_1m_2 + k_2m_1 - k_1m_2 - k_2l_1 = 0.$

We need to find out whether the conclusion polynomial z belongs to the radical of the ideal $(h_1, h_2, \ldots, h_6)$ or equivalently, whether 1 is an element of the ideal $I = (h_1, h_2, \ldots, h_6, zt - 1)$, where t is a slack variable. CoCoA returns

```
Use R::=Q[abcpqk[1..2]l[1..2]m[1..2]t];
I:=Ideal(p-l[1],ck[1]-bk[2],(p-k[1])b+(q-k[2])c,ac+bm[2]-cm[1]
-am[2],(p-m[1])(b-a)+(q-m[2])c,-acp+cp^2+abq-b^2q-c^2q+cq^2,
(l[1]m[2]+k[2]m[1]-k[1]m[2]-k[2]l[1])t-1);
NF(1,I);
```

the answer 1 and the statement is not generally true.

Let us look for non-degeneracy conditions. Elimination of dependent variables $p, q, k_1, k_2, l_1, m_1, m_2$ and t in the ideal I

```
Use R::=Q[abcpqk[1..2]l[1..2]m[1..2]t];
I:=Ideal(p-l[1],ck[1]-bk[2],(p-k[1])b+(q-k[2])c,ac+bm[2]-cm[1]
-am[2],(p-m[1])(b-a)+(q-m[2])c,-acp+cp^2+abq-b^2q-c^2q+cq^2,
(l[1]m[2]+k[2]m[1]-k[1]m[2]-k[2]l[1])t-1);
Elim(p..t,I);
```

gives the condition $(b^2 + c^2)((a - b)^2 + c^2) = 0$, which means that for the vertices of a triangle $B = C$ or $A = B$. We rule out these cases assuming that $B \neq C$ and $B \neq A$. We will add the polynomial $(b^2 + c^2)((a - b)^2 + c^2)v - 1$, where v is another slack variable, to the ideal I and the procedure now repeats. Denoting $J = I \cup \{(b^2 + c^2)((a - b)^2 + c^2)v - 1\}$ we get

```
Use R::=Q[abcpqk[1..2]l[1..2]m[1..2]vt];
J:=Ideal(p-l[1],ck[1]-bk[2],(p-k[1])b+(q-k[2])c,ac+bm[2]-cm[1]
-am[2],(p-m[1])(b-a)+(q-m[2])c,-acp+cp^2+abq-b^2q-c^2q+cq^2,
(b^2+c^2)((a-b)^2+c^2)v-1,(l[1]m[2]+k[2]m[1]-k[1]m[2]-k[2]l[1])
t-1);
NF(1,J);
```

the answer 0. The Simson–Wallace theorem is proved.

We showed that the condition "a point P lies on the circumcircle of ABC" is sufficient for the feet K, L, M to be collinear. Now we show that this condition is also necessary.

Suppose that the feet K, L, M from a point P onto the sides of ABC are collinear. We are to prove that P lies on the circumcircle of ABC. We ask if the polynomial $h_6 = -acp + cp^2 + abq - b^2q - c^2q + cq^2$ is an element of the radical of the ideal $(h_1, h_2, \ldots, h_5, z)$. Similarly, as in the previous case we find out non-degeneracy conditions $ac \neq 0$ assuming that $A \neq B$ and C does not lie on AB, which is obviously fulfilled in the case of a generic triangle. We add the polynomial $acv - 1$, where v is a slack variable, to the ideal $(h_1, h_2, \ldots, h_5, z, h_6t - 1)$. CoCoA returns

```
Use R::=Q[abcpqk[1..2]l[1..2]m[1..2]vt];
K:=Ideal(p-l[1],ck[1]-bk[2],(p-k[1])b+(q-k[2])c,ac+bm[2]-cm[1]
-am[2],(p-m[1])(b-a)+(q-m[2])c,l[1]m[2]+k[2]m[1]-k[1]m[2]-k[2]
l[1],acv-1,(-acp+cp^2+abq-b^2q-c^2q+cq^2)t-1);
NF(1,K);
```

the result `NF(1,K)=0` which means that the converse theorem holds as well. Hence, we can formulate the Simson–Wallace theorem in its stronger form [26]:

The locus of points P, whose orthogonal projections on the sides of a triangle ABC are collinear, is the circumcircle of ABC.

Now we give a classical proof of Simson–Wallace Theorem 4.1 (Fig. 4.3).

Let a point P be on the circumcircle of ABC. Denote K, L, M the feet of perpendiculars dropped from the point P to the sides BC, AC, AB respectively. We are to prove that K, L, M are collinear. We see, that it suffices to show the equality $\measuredangle CKL = \measuredangle BKM$.

Quadrilaterals $PCLK$ and $PKBM$ are cyclic, from which $\measuredangle CPL = \measuredangle CKL$ and $\measuredangle BPM = \measuredangle BKM$ follow. Hence we will show that $\measuredangle CPL = \measuredangle BPM$. Quadrilaterals $ABPC$ and $AMPL$ are cyclic as well. Since the sum of opposite angles in a cyclic quadrilateral equals π then $\measuredangle CPB = \pi - \measuredangle CAB$ and $\measuredangle LPM = \pi - \measuredangle CAB$ and from here $\measuredangle CPB = \measuredangle LPM$. Further $\measuredangle CPB = \measuredangle CPL + \measuredangle LPB$ and $\measuredangle LPM = \measuredangle LPB + \measuredangle BPM$, which gives the desired equality $\measuredangle CPL = \measuredangle BPM$.

Remark 4.1. Note that both proofs of the Simson–Wallace theorem — automatic and classical — required a certain ingenuity. In automatic proof

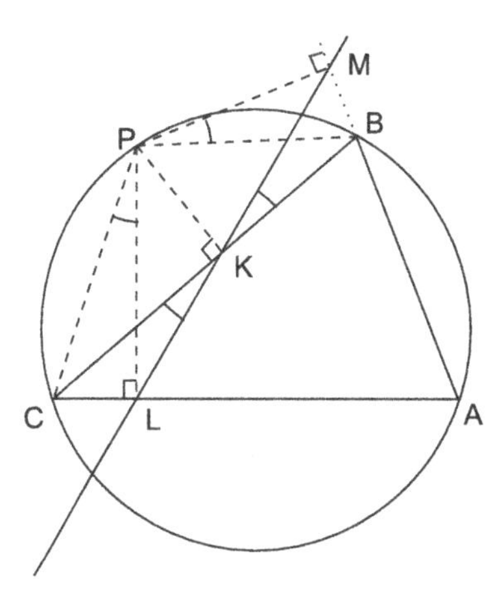

Fig. 4.3 Classical proof of the Simson–Wallace theorem

we needed to translate the condition "a point P lies on the circumcircle of ABC" into an algebraic equation, then to find non-degeneracy conditions and, essentially, to identify what non-degeneracy conditions mean geometrically. On the other hand in a classical proof at first we had to choose a strategy of proving the theorem, which need not be, in general, easy. A chosen strategy required special knowledge of the properties of cyclic quadrilaterals.

4.1 Gergonne's generalization

In the next part we will show a generalization of the Simson–Wallace theorem which is ascribed to J. D. Gergonne [22]. We will be concerned with the following problem:

Let K, L, M be the feet of perpendiculars dropped from a point P to the sides BC, CA, AB of a triangle ABC respectively. Instead of demanding K, L, M being collinear we look for points P leading to a pedal triangle KLM of fixed area s.

This problem is a generalization of the previous one since for zero area of KLM, that is, for the points K, L, M being collinear, the locus of points P is the circumcircle of ABC. To solve the problem we use the same Cartesian coordinate system and notation as in the last problem. Let $A = [a, 0]$, $B = [b, c]$, $C = [0, 0]$, $P = [p, q]$, $K = [k_1, k_2]$, $L = [l_1, 0]$, $M = [m_1, m_2]$

(Fig. 4.2). Suppose that the hypotheses $h_1, h_2, \ldots, h_5$ hold. For the area s of a triangle KLM we have

area of $KLM = s \Leftrightarrow h_7 : l_1m_2 + k_2m_1 - k_1m_2 - k_2l_1 - 2s = 0,$

since

$$s = \frac{1}{2} \begin{vmatrix} k_1 & k_2 & 1 \\ l_1 & 0 & 1 \\ m_1 & m_2 & 1 \end{vmatrix} . \tag{4.1}$$

Now the problem is more complex. Unlike the previous task we do not know the locus of points P — we have to *discover* it. First we try a validity of the statement, if no constraint is imposed on a point P. We find that the normal form of 1 in the ideal $(h_1, h_2, \ldots, h_5, h_7t - 1)$ equals 1, i.e., the statement is not proved, as we expected (we omit this part).

Further consider the ideal I which contains polynomials $h_1, h_2, \ldots, h_5$ and the condition h_7 of fixed area. In this ideal we eliminate all variables besides a, b, c, p, q, s. We get

```
Use R::=Q[abcpqk[1..2]l[1..2]m[1..2]s];
I:=Ideal(p-l[1],ck[1]-bk[2],(p-k[1])b+(q-k[2])c,ac+bm[2]-cm[1]
-am[2],(p-m[1])(b-a)+(q-m[2])c,l[1]m[2]+k[2]m[1]-k[1]m[2]-k[2]
l[1]-2s);
Elim(k[1]..m[2],I);
```

the equation of the form

$$C(s) : ac^3(p^2+q^2) - a^2c^3p + ac^2q(ab-b^2-c^2) + 2s(b^2+c^2)((a-b)^2+c^2) = 0.$$

A close inspection shows that the desired set is a circle $C(s)$ centered at $O = [q/2, (b^2 - ab + c^2)/(2c)]$ and radius

$$r = \sqrt{(b^2 + c^2)((a - b)^2 + c^2)(ac + 8s)/(4ac^3)} \tag{4.2}$$

which is concentric with the circumcircle of ABC (Fig. 4.4). We found that the necessary condition for the triangle KLM having fixed area s is, that a point P lies on the circle $C(s)$.

Conversely we will show that this condition is also sufficient. Does every point P of the circle $C(s)$ obey the condition that the triangle KLM has fixed area s? It suffices to explore if the polynomial h_7 belongs to the ideal $(h_1, h_2, \ldots, h_5, h_8)$, where

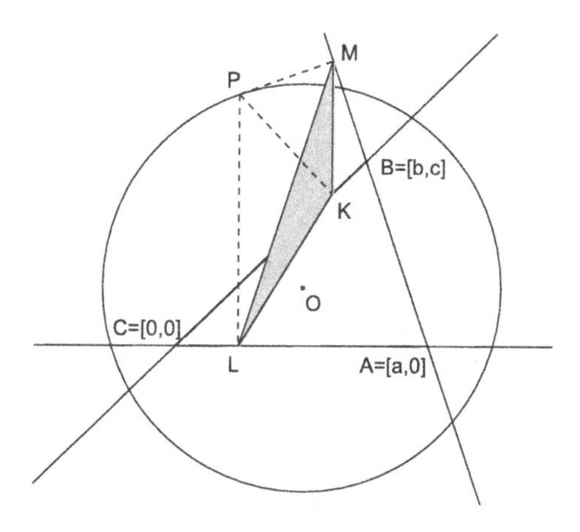

Fig. 4.4 Gergonne's generalization — triangle KLM has fixed area

$P \in C(s) \Leftrightarrow$

$h_8 : ac^3(p^2+q^2) - a^2c^3p + ac^2q(ab-b^2-c^2) + 2s(b^2+c^2)((a-b)^2+c^2) = 0.$

Adding non-degeneracy conditions $(b^2+c^2)((a-b)^2+c^2) \neq 0$, i.e., $B \neq C$ and $B \neq A$, to the ideal $(h_1, h_2, \ldots, h_5, h_8)$, CoCoA returns

```
Use R::=Q[abcpqk[1..2]l[1..2]m[1..2]svt];
J:=Ideal(p-l[1],ck[1]-bk[2],(p-k[1])b+(q-k[2])c,ac+bm[2]-cm[1]
-am[2],(p-m[1])(b-a)+(q-m[2])c,ac^3(p^2+q^2)-a^2c^3p+ac^2q(ab-
b^2-c^2)+2s(b^2+c^2)((a-b)^2+c^2),(b^2+c^2)((a-b)^2+c^2)v-1,
(l[1]m[2]+k[2]m[1]-k[1]m[2]-k[2]l[1]-2s)t-1);
NF(1,J);
```

the result `NF(1,J)=0` which means the validity of the converse statement. We rediscovered the following Gergonne's generalization of the Simson–Wallace theorem [22]:

Theorem 4.2 (Gergonne). *The feet of perpendiculars from P onto the sides of ABC form a triangle of the constant area s if and only if P lies on a circle which is concentric with the circumcircle of ABC with radius given by (4.2).*

Remark 4.2. We have to realize that we deal with the *signed* area. For instance, for $s = 2$ and $s = -2$ there exist *two* different concentric circles as geometric loci of points P, such that the triangles KLM and $K'L'M'$ related to the values $s = 2$ and $s = -2$ respectively, have the same (unsigned) area 2 but *opposite* orientation.

4.2 Generalization of Guzmán

The following generalization of the Simson–Wallace theorem was published in 1999 by M. de Guzmán [52]. He says that this generalization is easy to prove but it was not easy to discover it.

In the previous cases we projected a point P orthogonally to each side of a triangle ABC to obtain the points K, L, M. Now we will project a point P onto the sides BC, AC, AB of a triangle ABC in three *arbitrary* directions u, v, w given by vectors $\mathbf{u}$,$\mathbf{v}$,$\mathbf{w}$ to obtain the points K, L, M respectively. We exclude the case when all the three directions u, v, w are parallel (in this case points P fill the whole plane) and the case that directions u, v, w are parallel to the sides BC, AC, AB respectively. We will investigate the locus of points P such that the triangle KLM has constant area s (see [52]), where the problem is solved in a synthetic way.

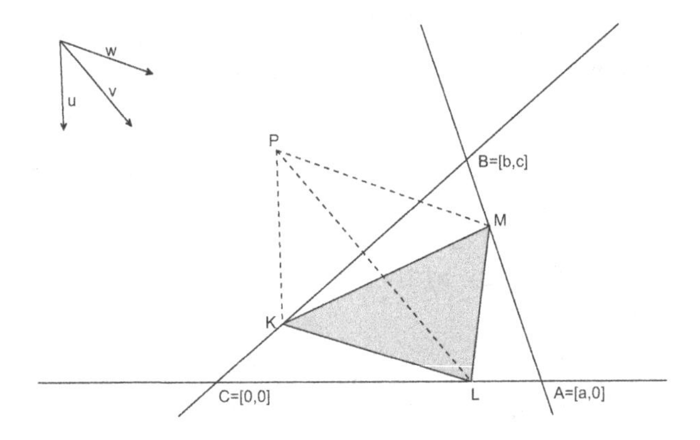

Fig. 4.5 Guzmán's generalization of Simson–Wallace theorem

Let us choose a Cartesian system of coordinates so that (Fig. 4.5), $A = [a, 0]$, $B = [b, c]$, $C = [0, 0]$, $K = [k_1, k_2]$, $L = [l_1, l_2]$, $M = [m_1, m_2]$, $P = [p, q]$, $\mathbf{u} = (u_1, u_2)$, $\mathbf{v} = (v_1, v_2)$, $\mathbf{w} = (w_1, w_2)$.

From $K = P + t_1\mathbf{u}$, $L = P + t_2\mathbf{v}$, $M = P + t_3\mathbf{w}$, $K = C + r_1(B - C)$, $L = C + r_2(A - C)$, $M = A + r_3(B - A)$, where $t_1, t_2, t_3, r_1, r_2, r_3$ are real parameters, we get the system of equations:

$K = P + t_1\mathbf{u} \Leftrightarrow h_1 : k_1 - p - t_1u_1 = 0,\ h_2 : k_2 - q - t_1u_2 = 0,$

$L = P + t_2\mathbf{v} \Leftrightarrow h_3 : l_1 - p - t_2v_1 = 0,\ h_4 : l_2 - q - t_2v_2 = 0,$

$M = P + t_3\mathbf{w} \Leftrightarrow h_5 : m_1 - p - t_3w_1 = 0,\ h_6 : m_2 - q - t_3w_2 = 0,$

$K \in BC \Leftrightarrow h_7 : k_1 - r_1b = 0,\ h_8 : k_2 - r_1c = 0,$

$L \in CA \Leftrightarrow h_9 : l_1 - r_2a = 0,\ h_{10} : l_2 = 0,$

$M \in AB \Leftrightarrow h_{11} : m_1 - a - r_3(b - a) = 0,\ h_{12} : m_2 - r_3c = 0.$

The conclusion h_{13} is by (4.1) of the form

area of $KLM = s \Leftrightarrow h_{13} : k_1l_2 + l_1m_2 + m_1k_2 - m_1l_2 - k_1m_2 - l_1k_2 - 2s = 0.$

It is obvious that, in general, h_{13} does not follow from the assumptions $h_1, h_2, \ldots, h_{12}$. Therefore we shall add the conclusion polynomial h_{13} to $h_1, h_2, \ldots, h_{12}$ and eliminate dependent variables $k_1, k_2, l_1, l_2, m_1, m_2, t_1, t_2, t_3, r_1, r_2, r_3$ in the ideal $I = (h_1, h_2, \ldots, h_{13})$ to find out desired additional conditions. We enter

```
Use R::=Q[abcpqsu[1..2]v[1..2]w[1..2]k[1..2]l[1..2]m[1..2]
t[1..3]r[1..3]];
I:=Ideal(k[1]-p-t[1]u[1],k[2]-q-t[1]u[2],l[1]-p-t[2]v[1],l[2]
-q-t[2]v[2],m[1]-p-t[3]w[1],m[2]-q-t[3]w[2],k[1]-r[1]b,k[2]-
r[1]c,l[1]-r[2]a,l[2],m[1]-a-r[3](b-a),m[2]-r[3]c,k[1]l[2]+
l[1]m[2]+m[1]k[2]-m[1]l[2]-k[1]m[2]-l[1]k[2]-2s);
Elim(k[1]..r[3],I);
```

and get an algebraic equation of the second degree in p, q

$$C(s) = 0, \tag{4.3}$$

where

$C(s) = c^2v_2p^2(u_1w_2 - u_2w_1) + cpq(au_2(v_1w_2 - v_2w_1) - (cv_1 + bv_2)(u_1w_2 - u_2w_1)) + cq^2(bv_1(u_1w_2 - u_2w_1) - au_1(v_1w_2 - v_2w_1)) - ac^2v_2p(u_1w_2 - u_2w_1) + acq(cu_1(v_1w_2 - v_2w_1) + bw_2(u_1v_2 - u_2v_1)) + 2v_2s(cu_1 - bu_2)(cw_1 + w_2(a - b)).$

As the constant s occurs only in the last term of (4.3) we can write

$$C(s) = C(0) + s \cdot Q, \tag{4.4}$$

where $Q = 2v_2(cu_1 - bu_2)(cw_1 + w_2(a - b))$. We have proved that (4.3) is a necessary condition for the area of KLM being s.

Now we will show that the condition (4.3) is sufficient as well. By the Hilbert's Nullstellensatz we are to prove that the polynomial h_{13} belongs

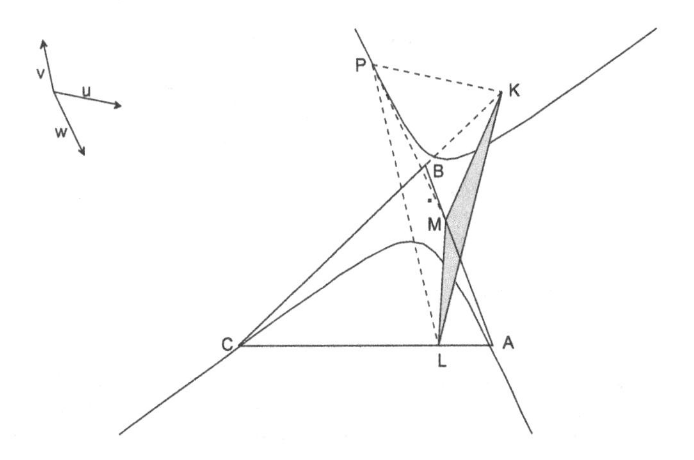

Fig. 4.6 For given directions u, v, w and $s \neq 0$ we get a hyperbola

to the radical ideal of $(h_1, h_2, \ldots, h_{12}, C(s))$ or equivalently, that 1 belongs to the ideal $J = (h_1, h_2, \ldots, h_{12}, C(s), h_{13}t - 1)$, where t is a slack variable. We compute the normal form `NF` of 1 with respect to the ideal J and get `NF(1,J)=1`, so it is not known if the condition (4.3) is sufficient for the area of KLM being s or not. The reason is that most geometric theorems are true generically.

We shall search for non-degeneracy conditions. Eliminating all dependent variables k_1, k_2, l_1, l_2, m_1, m_2, t_1, t_2, t_3, r_1, r_2, r_3 plus a slack variable t in the ideal J we get a condition

$$d : v_2(cu_1 - bu_2)(cw_1 + w_2(a - b)) = 0,$$

which means that at least one of the directions u, v, w is parallel to the sides BC, AC, AB respectively. To avoid this, we add the polynomial $dr - 1$ to the ideal J and compute the normal form of 1 with respect to the Gröbner basis of the ideal $J' = J \cup \{dr - 1\}$, where r is another slack variable. We obtain `NF(1,J')=0` and the condition (4.3) is sufficient for the area of KLM being s.

We arrived at the theorem which is due to M. de Guzmán [52]:

Theorem 4.3 (Guzmán). *Project P onto the sides BC, AC, AB of a triangle ABC in the given directions u, v, w which are not parallel to the sides BC, AC, AB into the points K, L, M respectively. We also exclude the case $u \parallel v \parallel w$. Then the locus of points P such that the area of a triangle K, L, M equals s is the conic $C(s)$ given by (4.3).*

We see that this generalization involves the previous cases. A demonstration of this generalization is carried out in dynamic geometry software Cabri II Plus (see Figs. 4.6 and 4.7).

Suppose that $c \neq 0$, $a \neq 0$, that is, A, B, C are not collinear and $A \neq B$. The family of conics $C(s)$ for a given s and arbitrary directions u, v, w has interesting properties which follow from (4.4). Let us recall some of them [52, 45]:

a) *$C(0)$ passes through the vertices A, B, C, i.e., $C(0)$ is a circumconic of*

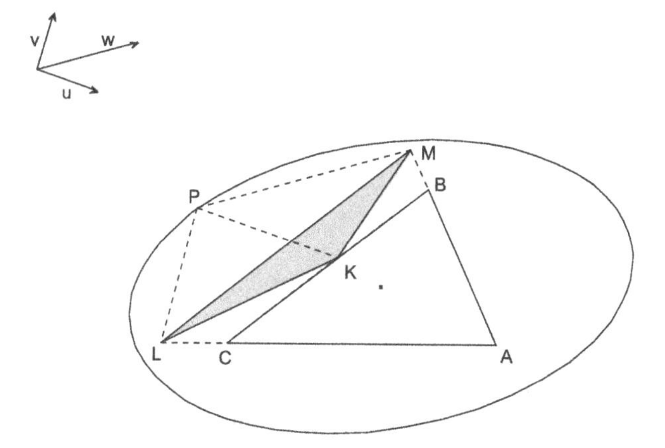

Fig. 4.7 For given directions u, v, w and $s \neq 0$ we get an ellipse

ABC.

The proof is obvious from the substitution of coordinates of A, B, C into the equation of $C(0)$. The same is seen from the fact that if a point P is at one vertex of ABC then for any directions u, v, w two vertices of K, L, M coincide with this vertex and points K, L, M are collinear.

b) *Varying the area s, the curves $C(s)$ to given directions u, v, w form a pencil of homothetic central conics (ellipses or hyperbolas with common axes) or a pencil of congruent parabolas with common axis.*

The equation $C(s) = 0$ implies that coefficients of quadratic terms simultaneously vanish iff the vectors **u**,**v**,**w** are dependent, i.e., iff the directions u, v, w are pairwise parallel which we excluded. Since the constant s occurs only in the absolute term of the equation $C(s)$ then the conic $C(s)$ has the same asymptotic directions for an arbitrary s.

c) $C(0)$ *is singular if and only if two of the directions* u, v, w *are parallel.*

The determinant of $C(0)$ vanishes iff

$$(v_1w_2 - v_2w_1)(u_1v_2 - u_2v_1)(u_1w_2 - u_2w_1) = 0,$$

from which our statement follows. If, for example, $u \parallel v$ then a conic consists of two straight lines — the straight line AB and a straight line passing through the vertex C in the direction $u \parallel v$. If $u \parallel v \parallel AB$ then we get two parallels (which coincide for $s = 1/2$).

d) *If all the directions* u, v, w *are pairwise different then* $C(s)$ *is a regular conic.*

This follows from c).

e) $C(0)$ *passes through the points* S, T, U*, where* $S = SB \cap SC$, $T = TA \cap TC$ *and* $U = UA \cap UB$*, where* $SB \parallel TA \parallel w$, $SC \parallel UA \parallel v$ *and* $TC \parallel UB \parallel u$.

We will see this later (Fig. 4.8).

f) *If we choose directions* $u = (c, -b)$, $v = (0, 1)$, $w = (c, a - b)$ *then the conic* $C(0)$ *is a circle*

$$cp^2 + cq^2 - acp + abq - b^2q - c^2q = 0.$$

The proof follows from a substitution of vector coordinates into the equation $C(0) = 0$.

In Fig. 4.6 a hyperbola for directions u, v, w and $s \neq 0$ is depicted. Further on we see an ellipse for directions u, v, w and $s \neq 0$ in Fig. 4.7.

Now we will show how we can construct a conic $C(0)$ by a compass and ruler for given directions u, v, w, [52] (Fig. 4.8).

Through the vertices A, B of a triangle ABC we lead parallels to directions v, u respectively. Denote their intersection by U. Similarly we construct parallels w, v through B, C and parallels u, w through the vertices C, A. Denote their intersections by S, T. The points S, T, U belong to the conic $C(0)$ as we can easily see. We get six points which form a hexagon $AUBSCT$ that is inscribed into the conic and whose opposite sides are parallel. Centers of the opposite sides lie on a diameter of the conic, for instance, the centers of AU and SC lie on a diameter which, as known, passes through the center of the conic. All three diameters intersect at the center of the conic (if it exists). If diameters are parallel then they are parallel to the axis of a parabola. By Pascal's theorem we can construct more points of the conic.

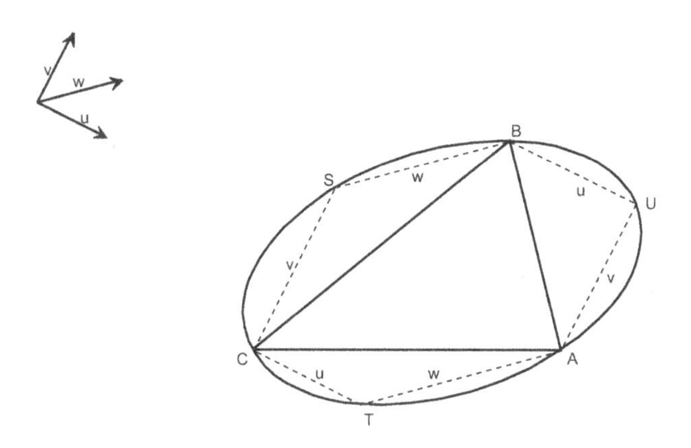

Fig. 4.8 Construction of a conic $C(0)$ with given directions u, v, w

Remark 4.3.

1) In the paper [45] affine and projective generalization of the Simson–Wallace theorem are introduced. If the affine feet K, L, M lie on the affine Wallace line of P with respect to a center Z or if the projective feet K, L, M lie on the projective Simson–Wallace line of P with respect to a center Z and an axis z, then P lies on a conic which depends on two parameters given by coordinates of the center Z (see also [107]).

2) Given a triangle ABC, the conic (4.3) is defined by seven parameters $u_1, u_2, v_1, v_2, w_1, w_2$ and s, but in fact four parameters are enough by setting $u_2 = v_2 = w_2 = 1$ (besides the cases $u_2 = 0, v_2 = 0, w_2 = 0$). Briefly we can write $C(u, v, w, s)$.

3) In the classical Simson–Wallace theorem the directions u, v, w and the directions of the sides of the given triangle form three pairs of an involutoric projectivity, which includes the so-called "absolute involution ι" in the ideal line of the plane. As any circle in plane the circles of Gergonne's extension of the Simson–Wallace theorem pass through the imaginary fixed points of ι. Choosing u, v, w arbitrarily, the above mentioned pairs of directions define an elliptic or hyperbolic or parabolic projectivity π within the ideal line of the plane and the solution conics $C(s)$ of Guzmán's generalization pass through the fixed points of this projectivity π. As a consequence, triplets (u_i, v_i, w_i) defining projectivities π_i with common fixed points lead to identical sets of conics $\{C_i(s)\}$.

4.3 Generalization to three dimensions

In this part we generalize the Simson–Wallace theorem to the space. We will show two generalizations. First we extend the Simson–Wallace theorem to the space considering a tetrahedron $ABCD$ instead of a triangle and orthogonal projections K, L, M, N of a point P onto the faces of $ABCD$ such that the volume of $KLMN$ equals s.

Then we will replace a tetrahedron by a skew quadrilateral and investigate the similar problem.

4.3.1 *Generalization on a tetrahedron*

Consider a tetrahedron $ABCD$ in the Euclidean space E^3. Let P be an arbitrary point and K, L, M, N the feet of perpendiculars dropped from P onto the faces BCD, ACD, ABD, ABC of the tetrahedron $ABCD$ respectively. We are looking for a locus of points P such that the volume of the tetrahedron $KLMN$ equals the constant s. The generalization for a tetrahedron has also been studied by F. Botana [12] and E. Roanes–Lozano, [111].

Choose a Cartesian system of coordinates so that $A = [0,0,0]$, $B = [a,0,0]$, $C = [b,c,0]$, $D = [d,e,f]$, $K = [k_1,k_2,k_3]$, $L = [l_1,l_2,l_3]$, $M = [m_1,m_2,m_3]$, $N = [n_1,n_2,n_3]$, $P = [p,q,r]$ (Fig. 4.9).

Then the following relations hold:

$PK \perp BCD \Leftrightarrow$
$h_1 : (b-a)(p-k_1)+c(q-k_2) = 0,\ h_2 : (d-a)(p-k_1)+e(q-k_2)+f(r-k_3) = 0,$

$K \in BCD \Leftrightarrow$
$h_3 : -acf - aek_3 + afk_2 + ack_3 + cfk_1 + bek_3 - cdk_3 - bfk_2 = 0,$

$PL \perp ACD \Leftrightarrow$
$h_4 : b(p-l_1) + c(q-l_2) = 0,\ h_5 : d(p-l_1) + e(q-l_2) + f(r-l_3) = 0,$

$L \in ACD \Leftrightarrow$
$h_6 : cfl_1 + bel_3 - cdl_3 - bfl_2 = 0,$

$PM \perp ABD \Leftrightarrow$
$h_7 : a(p-m_1) = 0,\ h_8 : d(p-m_1) + e(q-m_2) + f(r-m_3) = 0,$

$M \in ABD \Leftrightarrow$
$h_9 : aem_3 - afm_2 = 0,$

$PN \perp ABC \;\Leftrightarrow$
$h_{10} : a(p - n_1) = 0,\; h_{11} : b(p - n_1) + c(q - n_2) = 0,$

$N \in ABC \Leftrightarrow$
$h_{12} : acn_3 = 0.$

Conclusion h_{13} is of the form:

$$\text{volume of } KLMN = s \quad \Leftrightarrow \quad h_{13} : \begin{vmatrix} k_1 & k_2 & k_3 & 1 \\ l_1 & l_2 & l_3 & 1 \\ m_1 & m_2 & m_3 & 1 \\ n_1 & n_2 & n_3 & 1 \end{vmatrix} - 6s = 0. \qquad (4.5)$$

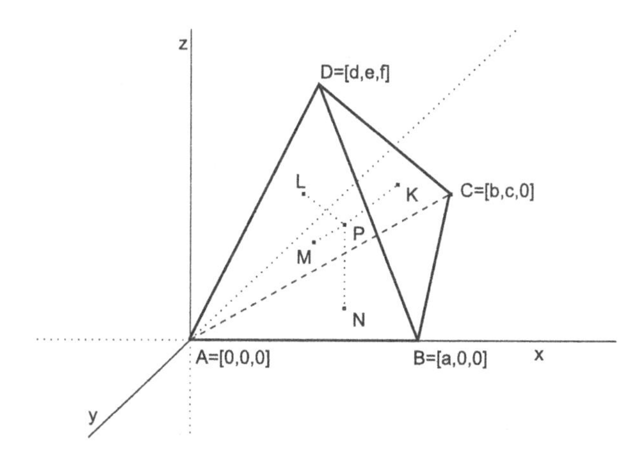

Fig. 4.9 Generalization of Simson–Wallace theorem on a tetrahedron

The direct elimination of dependent variables k_1, k_2, k_3, l_1, l_2, l_3, m_1, m_2, m_3, n_1, n_2, n_3 in the ideal $I = (h_1, h_2, \ldots, h_{12}, h_{13})$ fails. Hence we use the following *successive* elimination.

First we eliminate $m_1, m_2, m_3, n_1, n_2, n_3$ in the ideal $(h_7, \ldots, h_{13})$. We obtain the elimination ideal generated by the polynomial p_1. Then eliminate l_1, l_2, l_3 in the ideal (h_4, h_5, h_6, p_1). We get the elimination ideal with one generator p_2. In the end we eliminate k_1, k_2, k_3 in the ideal (h_1, h_2, h_3, p_2) to obtain the only condition

$$F(s) := ac^2 f^3 G + s \cdot Q = 0, \qquad (4.6)$$

where

$G = bf^2q^3(b-a) + fr^3(abe - acd + cd^2 - b^2e - c^2e + ce^2) + c^2f^2p^2q + cfp^2r(e^2 - ce + f^2) + cf^2q^2p(a-2b) + fq^2r(abe - acd + cd^2 - b^2e + cf^2) + cf^2r^2p(a-2d) + f^2r^2q(b^2 - ab + c^2 - 2ce) + 2cefpqr(b-d) + abcf^2q^2 + r^2(abce^2 - ac^2de + c^2d^2e + acde^2 - 2bcde^2 - abe^3 + b^2e^3 + acdf^2 - abef^2 + b^2ef^2 + c^2ef^2) - ac^2f^2pq + acfpr(ce - e^2 - f^2) + fqr(ac^2d - 2abce - c^2d^2 + 2bcde - b^2e^2 + abe^2 + abf^2 - b^2f^2 - c^2f^2)$

and

$Q = -6(e^2+f^2)((cd-be)^2 + b^2f^2 + c^2f^2)(a^2c^2 - 2ac^2d + c^2d^2 - 2a^2ce + 2abce + 2acde - 2bcde + a^2e^2 - 2abe^2 + b^2e^2 + a^2f^2 - 2abf^2 + b^2f^2 + c^2f^2)$.

We see that Q is a constant which does not depend on p, q, r, s.

We established the following theorem [89]:

Theorem 4.4. *Let K, L, M, N be orthogonal projections of an arbitrary point P consecutively on the faces BCD, ACD, ABD, ABC of a tetrahedron $ABCD$. Then the points P such that the tetrahedron $KLMN$ has constant volume s belong to the surface $F(s) = 0$ from (4.6).*

Remark 4.4.

1) Q can be written in the form

$$Q = -6\cdot\frac{1}{a^2}\cdot|(B-A)\times(D-A)|^2\cdot|(D-A)\times(C-A)|^2\cdot|(D-B)\times(C-B)|^2. \tag{4.7}$$

2) Cf. [111], where the same problem is solved for a special class of tetrahedra (in our notation) $A = [0,0,0], B = [1,0,0], C = [0,c,0], D = [1,e,f]$.

3) Note that (4.6) is a necessary condition for the tetrahedron $KLMN$ to have constant volume. We did not succeed to show that (4.6) is (in this general form) also sufficient for the moment. Hence, we do not know whether every point of the surface $F(s) = 0$ obeys the conditions of the theorem. In [111] a similar problem for a special tetrahedron is solved by Wu's method [144], which is based on pseudodivision. In this way a necessary and sufficient condition was found in accordance with our results. We will show that for concrete values a, b, c, d, e, f we are able to check sufficiency as well.

Now we give some properties of a surface $F(s) = 0$ for s being zero, i.e., when K, L, M, N are complanar. From (4.6) $F(0) = 0 \Leftrightarrow G = 0$ follows.

Theorem 4.5. *The surface G has the following properties* [89]*:*

a) G contains the edges AB, AC, AD, BC, BD, CD of $ABCD$, that is, G is

a circumsurface of $ABCD$.
b) G is a cubic surface.
c) G has 4 singular points — the vertices A, B, C, D of the tetrahedron.
d) The point which is in the intersection of three planes which pass through the edges AB, BD, DA and are perpendicular to the planes ABC, BDC, DAC belongs to the surface G. Similarly we will proceed for another triple of edges.
e) The lines AB, AC, AD, BC, BD, CD are torsal lines of the cubic G, i.e., the tangent plane at an arbitrary point of the torsal line contains the whole line. The tangent planes at three pairs of opposite edges intersect at another 3 straight lines which are complanar. Each of these three lines intersects the pair of skew torsal lines.
f) There exists a simple rational parametrization of G.

Proof.
a) Realize that an arbitrary point P of an edge coincides with *two* points from the feet of perpendiculars K, L, M, N, which are then complanar. Another way to verify this is a direct computation.
b) If all cubic terms in the equation $G = 0$ vanish then the surface is a quadric, which is not possible, because the surface contains all six edges of $ABCD$.
c) The statement follows from the fact, that the cubic surface contains all the six edges of a tetrahedron $ABCD$ (see [20]).
d) To prove this realize that in this case the feet K, L, M, N lie in the plane ABD.
e) For the proof see [20], pp. 567–568.
f) Let $X = A + t\mathbf{u}$ be a straight line with $\mathbf{u} = (u, v, 1)$. Then, since A is a double point, it intersects the cubic surface G at most at one point $X(u, v, 1)$ (see Example 4.1 below). □

Remark 4.5.
1) It is well-known that every cubic surface (in the complex projective space) contains 27 lines. In this case we have $6 \times 4 + 3 = 27$ lines, because each edge is counted four times, see [20, 55, 95].

2) To prove Theorem 4.4 we could also proceed in a more synthetic way similarly as Guzmán in the planar case [52].
The feet K, L, M, N form a tetrahedron of fixed volume s. Hence K, L, M, N fulfil the formula (4.5). The coordinates of points K, L, M, N being intersections of perpendiculars from $P = [p, q, r]$ to the faces of

$ABCD$ are *linear* in p, q, r. Thus (4.5) is a cubic algebraic equation in p, q, r, which has in general 20 real coefficients. To determine these coefficients we need at least 19 points of the surface. We know that each edge of $ABCD$ contains two double points, which makes together $4 + 6 \times 2 = 16$ points. The last three points remain to be determined. Construction of such points follows from d) of Theorem 4.5.

Example 4.1. For special values $a = 1$, $b = 0$, $c = 1$, $d = 0$, $e = 0$, $f = 1$, $s = 0$ we get from (4.6) the equation of a cubic surface

$$p^2q + pq^2 + p^2r + q^2r + pr^2 + qr^2 - pq - pr - qr = 0 \tag{4.8}$$

(see Fig. 4.10).

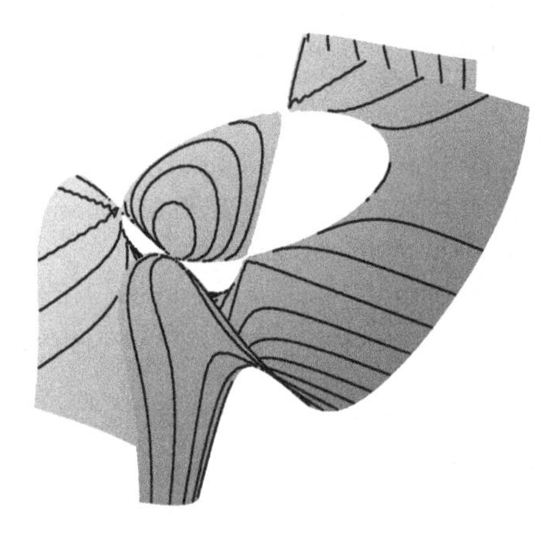

Fig. 4.10 Cubic surface $p^2q + pq^2 + p^2r + q^2r + pr^2 + qr^2 - pq - pr - qr = 0$ as the locus of points P with complanar feet of a (special) tetrahedron

First we will prove that (4.8) is also a sufficient condition for K, L, M, N being complanar (see Example 4.4), where detailed computation is carried out.

This surface can be rationally parametrized easily, taking into account that the surface has 4 double points. Putting $p = ur, q = vr, r = r$ and setting this into (4.8) we get

$$\begin{aligned} p &= \frac{u(u + uv + v)}{u^2v + u^2 + uv^2 + v^2 + u + v} \\ q &= \frac{v(u + uv + v)}{u^2v + u^2 + uv^2 + v^2 + u + v} \\ r &= \frac{u + uv + v}{u^2v + u^2 + uv^2 + v^2 + u + v} \end{aligned} \tag{4.9}$$

for real parameters u, v.

Example 4.2. The choice $a = 2$, $b = 1$, $c = \sqrt{3}$, $d = 1$, $e = 1/\sqrt{3}$, $f = \sqrt{8/3}$ with the centroid of $ABCD$ in the origin for an arbitrary s gives a one-parametric system of surfaces which are associated with a regular tetrahedron (writing x, y, z instead of p, q, r):

$$24\sqrt{6}x^2y + 24\sqrt{3}x^2z + 24\sqrt{3}y^2z - 8\sqrt{6}y^3 - 16\sqrt{3}z^3 + 36\sqrt{2}x^2 + 36\sqrt{2}y^2 \\ +36\sqrt{2}z^2 - 18\sqrt{2} - 729s = 0. \tag{4.10}$$

For $s = 0$ we obtain the locus of points P such that the feet K, L, M, N are complanar. This cubic surface has the vertices A, B, C, D as the only singular points.

For $s \neq 0$ all the cubic surfaces associated with a regular tetrahedron $ABCD$ do not contain singular points unless the value $s = -18\sqrt{2}/729$. This leads to the cubic surface

$$6\sqrt{6}x^2y + 6\sqrt{3}x^2z + 6\sqrt{3}y^2z - 2\sqrt{6}y^3 - 4\sqrt{3}z^3 + 9\sqrt{2}x^2 + 9\sqrt{2}y^2 + 9\sqrt{2}z^2 = 0, \tag{4.11}$$

with one singular point — an isolated point placed in the centroid of $ABCD$.

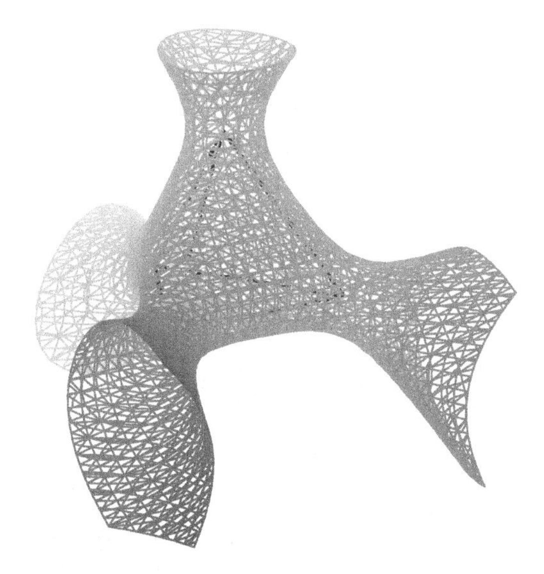

Fig. 4.11 Cubic surface associated with a regular tetrahedron as the locus of points P with constant volume $s = 10\sqrt{2}/729$ of a tetrahedron $KLMN$.

In Fig. 4.11 we see a cubic surface (4.10) associated with a regular tetrahedron $ABCD$ for $s = 10\sqrt{2}/729$.

4.3.2 *Generalization on a skew quadrilateral*

Another generalization of the Simson–Wallace theorem is based on a consideration of a skew quadrilateral $ABCD$ in E^3 instead of a tetrahedron. Denote by K, L, M, N the feet of perpendiculars which are dropped from a point P onto the sides AB, BC, CD, DA of a quadrilateral $ABCD$ respectively. We are to find the locus of points P such that the feet K, L, M, N are complanar.

Choose a Cartesian system of coordinates so that $A = [0,0,0]$, $B = [a,0,0]$, $C = [b,c,0]$, $D = [d,e,f]$, $P = [p,q,r]$, $K = [k_1,0,0]$, $L = [l_1,l_2,0]$, $M = [m_1,m_2,m_3]$, $N = [n_1,n_2,n_3]$ (Fig. 4.12). The

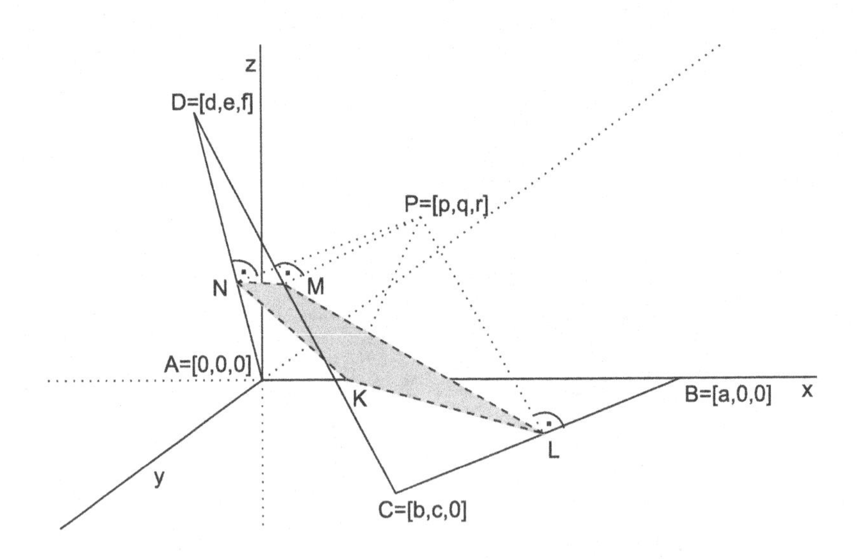

Fig. 4.12 Generalization of Simson–Wallace theorem on a skew quadrilateral

conditions are as follows:

$$
\begin{aligned}
PK \perp AB \Leftrightarrow\ & h_1 : a(p-k_1)=0,\\
L \in BC \Leftrightarrow\ & h_2 : l_2(b-a)-c(l_1-a)=0,\\
PL \perp BC \Leftrightarrow\ & h_3 : (p-l_1)(b-a)+c(q-l_2)=0,\\
M \in CD \Leftrightarrow\ & h_4 : (d-b)(m_2-c)-(e-c)(m_1-b)=0,\\
& h_5 : (e-c)m_3-(m_2-c)f=0,\\
& h_6 : (m_1-b)f-m_3(d-b)=0,\\
PM \perp CD \Leftrightarrow\ & h_7 : (p-m_1)(d-b)+(q-m_2)(e-c)+(r-m_3)f=0,\\
N \in DA \Leftrightarrow\ & h_8 : dn_2-en_1=0,\\
& h_9 : dn_3-fn_1=0,\\
& h_{10} : fn_2-en_3=0,\\
PN \perp DA \Leftrightarrow\ & h_{11} : (p-n_1)d+(q-n_2)e+(r-n_3)f=0.
\end{aligned}
$$

By (4.5) the last condition h_{12} has the form:

K, L, M, N are complanar $\Leftrightarrow$

$h_{12} : k_1l_2m_3 - l_2m_3n_1 - k_1m_3n_2 + l_1m_3n_2 - k_1l_2n_3 + l_2m_1n_3 + k_1m_2n_3 - l_1m_2n_3 = 0.$

The successive elimination (first eliminate n_1, n_2, n_3, then m_1, m_2, m_3 etc.) of the 9 variables $k_1, \ldots, n_3$ from the ideal $(h_1, h_2, \ldots, h_{11}, h_{12})$ gives the following equation $H = 0$ of a cubic surface $H(p, q, r)$ which contains 176 terms:

$H := p^3(c^2d(-a+d)-(be^2+bf^2-2cde)(a-b))+p^2qc(-ace+ae^2-af^2+2bf^2)+p^2rcf(-ac+2cd+2ae-2be)+pq^2(c^2(-ad+d^2+f^2)+(2cde-be^2)(a-b))+pqr2f(cd-be)(a-b)+pr^2f^2(-ab+b^2+c^2)+q^3ace(-c+e)+q^2racf(-c+2e)+qr^2acf^2+p^2(cd(a^2c-cd^2-2a^2e+abe+b^2e+c^2e-ade+bde)+(e^2+f^2)(a^2b-b^3+ac^2-bc^2-c^2d-ace+bce))+pq(-a^2bcd+ab^2cd+ac^3d+a^2cd^2-b^2cd^2-c^3d^2+a^2c^2e-a^2ce^2-de(b^2+c^2)(a-b)+f^2(a^2c-abc-b^2c-c^3)-(d^2+e^2+f^2)(cd-be)(a-b))+prf((-ab+b^2+c^2)(bd+ce-d^2-e^2-f^2)+ac(2be+ac-2cd-2ae))+q^2ae(c(bd+ce-d^2-e^2-f^2)+(-c+e)(ab-b^2-c^2))+qra(cf(bd+ce-d^2-e^2-f^2)+f(-c+2e)(ab-b^2-c^2))+r^2af^2(ab-b^2-c^2)-pa(cd(acd-cd^2+c^2e-(be+de)(a-b))+(e^2+f^2)((b^2+c^2-ce)(a-b)-c^2d))+(qe+rf)a(bd+ce-d^2-e^2-f^2)(ab-b^2-c^2) = 0.$

The equation $H = 0$ gives a necessary condition for $P = [p, q, r]$ such that the feet K, L, M, N are complanar. We can state a theorem [89]:

Theorem 4.6. *Let P be an arbitrary point and K, L, M, N the feet of perpendiculars dropped from P onto the sides AB, BC, CD, DA of a skew quadrilateral $ABCD$ respectively. Then the point $P = [p, q, r]$ such that K, L, M, N are complanar obeys the equation $H = 0$.*

Remark 4.6. Proving that the condition $H = 0$ (in this general form) is also sufficient for K, L, M, N being complanar failed for the moment. In concrete cases (as in the next example) this verification has been done.

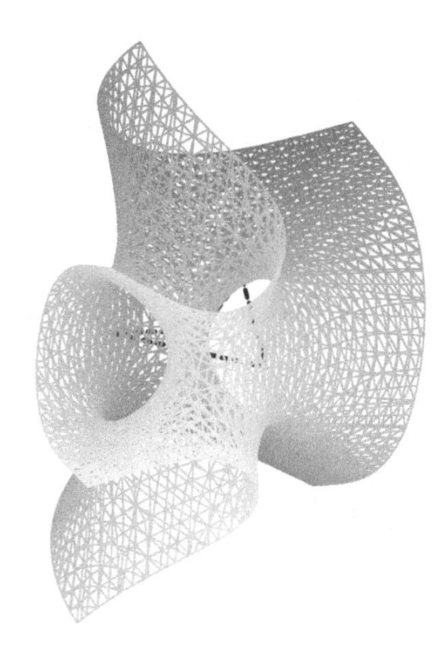

Fig. 4.13 Cubic surface $p^2q + pq^2 - p^2r - q^2r + pr^2 + qr^2 - 2pq - r^2 + r = 0$ as the locus of points P with complanar feet on the edges of a skew quadrilateral

Let us see the example

Example 4.3. The choice $a = 1$, $b = 1$, $c = 1$, $d = 0$, $e = 0$, $f = 1$ leads to the cubic surface

$$h := p^2q + pq^2 - p^2r - q^2r + pr^2 + qr^2 - 2pq - r^2 + r = 0. \tag{4.12}$$

The verification that (4.12) is also sufficient for K, L, M, N being complanar is as follows:

```
Use R::=Q[abcdefpqrk[1..3]l[1..3]m[1..3]n[1..3]t];
I:=Ideal(a(p-k[1]),l[2](b-a)-c(l[1]-a),(p-l[1])(b-a)+c(q-l[2]),
```

```
(d-b)(m[2]-c)-(e-c)(m[1]-b),(e-c)m[3]-(m[2]-c)f,(m[1]-b)f-
m[3](d-b),(p-m[1])(d-b)+(q-m[2])(e-c)+(r-m[3])f,dn[2]-en[1],
dn[3]-fn[1],fn[2]-en[3],(p-n[1])d+(q-n[2])e+(r-n[3])f,a-1,b-1,
c-1,d,e,f-1,p^2q+pq^2-p^2r-q^2r+pr^2+qr^2-2pq-r^2+r,
(k[1]l[2]m[3]-l[2]m[3]n[1]-k[1]m[3]n[2]+l[1]m[3]n[2]-
k[1]l[2]n[3]+l[2]m[1]n[3]+k[1]m[2]n[3]-l[1]m[2]n[3])t-1);
NF(1,I);
```

The answer `NF=0` immediately follows.

Investigating singular points of (4.12) gives the system of equations

$$\begin{aligned} h &:= p^2q + pq^2 - p^2r - q^2r + pr^2 + qr^2 - 2pq - r^2 + r = 0, \\ \frac{\partial h}{\partial p} &:= 2pq + q^2 - 2pr + r^2 - 2q = 0, \\ \frac{\partial h}{\partial q} &:= p^2 + 2pq - 2qr + r^2 - 2p = 0, \\ \frac{\partial h}{\partial r} &:= -p^2 - q^2 + 2pr + 2qr - 2r + 1 = 0. \end{aligned} \tag{4.13}$$

Solving the system (4.13) we eliminate unknowns q, r in the ideal $J = (h, \partial h/\partial p, \partial h/\partial q, \partial h/\partial r)$ to get p, then p, r to get q and in the end p, q to obtain r. In CoCoA we enter

```
Use R::=Q[pqr];
J:=Ideal(p^2q+pq^2-p^2r-q^2r+pr^2+qr^2-2pq-r^2+r,2pq+q^2-2pr
-2qr+r^2-2q,p^2+2pq-2qr+r^2-2p,-p^2-q^2+2pr+2qr-2r+1);
Elim(q..r,J);
```

with response

```
Ideal(-19/12)
```

We get similar answers for unknowns q and r. Hence, the cubic (4.12) does not have any singular points. The cubic (4.12) is depicted in Fig. 4.13.

Let us look at another case

Example 4.4. The choice $a = 1$, $b = 0$, $c = 1$, $d = 0$, $e = 0$, $f = 1$ leads to the equation

$$-p^2q + pq^2 - p^2r - q^2r + pr^2 + qr^2 + p^2 - r^2 - p + r = 0, \tag{4.14}$$

or after factorization

$$(p - r)(pq - q^2 + pr + qr - p - r + 1) = 0.$$

Thus the cubic surface (4.14) decomposes into the plane and one-sheet hyperboloid (see Fig. 4.14).

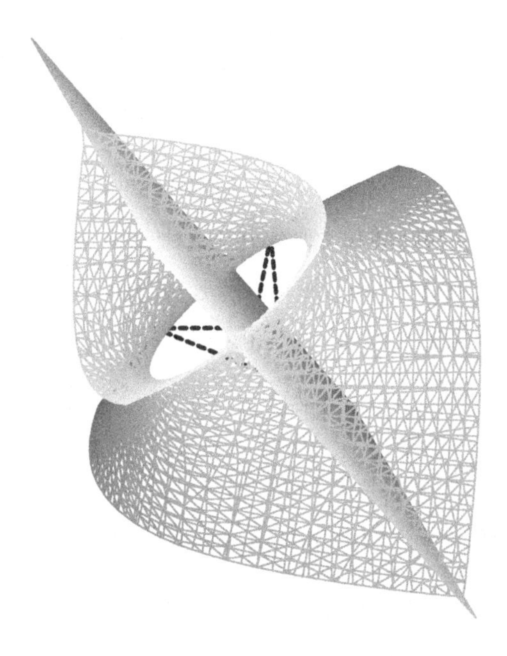

Fig. 4.14 Plane and quadric $(p-r)(pq-q^2+pr+qr-p-r+1)=0$ as the locus of points P with complanar feet on the edges of a skew quadrilateral

Remark 4.7.

1) The generalization above stimulates immediately the following question: The 6 edges of a tetrahedron allow three possibilities of skew edge quadrilaterals. How are the three solution surfaces related to these three possibilities (Fig. 4.14)?

2) For the construction of the generalization above it is not essential that the four edges form a skew quadrilateral. One could equally treat the case of four skew given lines a, b, c, d and ask for complanar pedal points K, L, M, N of a point P. And now it would be interesting to know if it makes a difference whether the given lines are generators of a regulus or not [138].

3) In Fig. 4.14 (Example 4.4) the respective cubic surface decomposes into a quadric and a plane. In other cases (Example 4.3) we get an irreducible

cubic. Why?

The Simson–Wallace theorem has been generalized several times throughout history. Two generalizations to three dimensions are based on results of commutative algebra in the last third of the last century. There are many questions arising from this, with some of them being indicated in remarks. There are problems with computational complexity. The equations of surfaces are too long in its general form, which prevents us from other generalizations (arbitrary directions u, v, w, investigation of the volume of a quadrilateral $KLMN$ in the second generalization, etc.).

Chapter 5

Transversals in a polygon

In many tasks of elementary geometry, properties of transversals of polygons like altitudes, medians, side and angle bisectors, etc. are investigated. Theorems of Ceva and Menelaus are the best known among theorems about properties of transversals. Other properties of transversals are explored in the theorem of Euler.

In this chapter we will be concerned with the above theorems and their generalizations both in the plane and space. The properties of transversals will be studied by means of computer using the theory of automatic theorem proving. By this method we are able to derive most relations that are already known so far. The problems are also solved by a classical way so as to show the strengths and weaknesses of both computer and classical methods. In the classical approach we will use the powerful area principle [49] (or the area method) which is based on ratios of the areas of triangles. In the following text we try to gather problems of this type.

Investigation of transversals of a polygon by computer requires one to describe analytically geometric properties of an object. Usually we introduce a Cartesian system of coordinates to carry out this algebraic translation. In some instances we can use a simpler *affine* system of coordinates (sometimes the words *linear* system are used instead) since we will mostly study geometric properties which preserve the ratio of the lengths of segments or the ratio of the areas of triangles, which we call affine properties and affine invariants. Using an affine coordinate system we usually "save" at least one variable in the plane, two variables in the three dimensional space, etc. in comparison with the use of a Cartesian coordinate system. We will appreciate this especially when computation is too complex and we have many variables.

5.1 Theorem of Ceva

In this section we will study the well-known theorem of Ceva. It was established in 1678 by the Italian mathematician Giovanni Ceva (1648–1734). We will formulate Ceva's theorem in terms of ratios of points. Let us recall the definition of the ratio of a point [96, 120].

Definition 5.1. Let A, B, C be three distinct collinear points in an affine space. Then the *ratio of a point C with respect to the points A, B* is a real number λ which satisfies

$$C - A = \lambda(C - B), \tag{5.1}$$

where $C - A, C - B$ are vectors $\overrightarrow{AC}, \overrightarrow{BC}$ respectively. We will denote $(ABC) = \lambda$.

We can equivalently say that

$$|(ABC)| = \frac{|AC|}{|BC|}, \tag{5.2}$$

and (ABC) is negative if C lies between A and B, and is positive otherwise.

The theorem of Ceva reads:

Theorem 5.1 (Ceva). *Given a triangle ABC and three points A', B', C' on the lines BC, AC, AB. Then the lines AA', BB', CC' are concurrent or parallel if and only if*

$$(ABC') \cdot (BCA') \cdot (CAB') = -1. \tag{5.3}$$

First we will derive (5.3) by computer.

Denote by $l_1 = (ABC')$ the ratio of the point C' with respect to the points A, B and analogously $l_2 = (BCA')$ and $l_3 = (CAB')$. By (5.1) we can write

$$\begin{aligned} l_1 &= (ABC') \Leftrightarrow l_1(C' - B) = C' - A, \\ l_2 &= (BCA') \Leftrightarrow l_2(A' - C) = A' - B, \\ l_3 &= (CAB') \Leftrightarrow l_3(B' - A) = B' - C. \end{aligned} \tag{5.4}$$

Choose an affine coordinate system so that $A = [0, 0]$, $B = [a, 0]$, $C = [0, c]$, $A' = [p, q]$, $B' = [0, r]$, $C' = [s, 0]$ (Fig. 5.1). From the relations (5.4) have the following equations:

$l_1 = (ABC') \Leftrightarrow h_1 : l_1(s - a) - s = 0,$

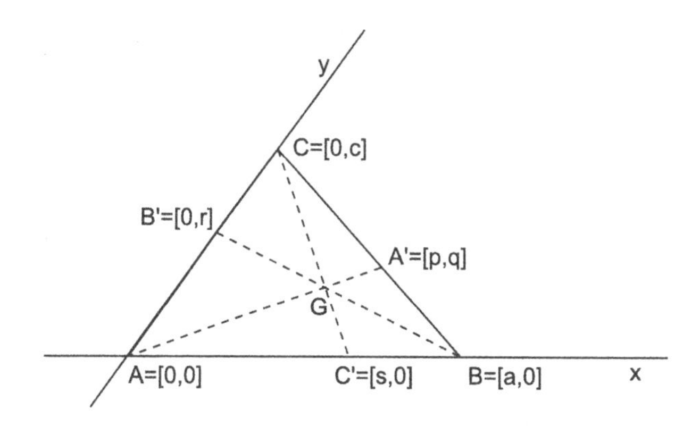

Fig. 5.1 Ceva's theorem: The lines AA', BB', CC' are concurrent $\Leftrightarrow (ABC')\cdot(BCA')\cdot(CAB') = -1$

$$l_2 = (BCA') \Leftrightarrow h_2 : l_2p - p + a = 0,$$

$$h_3 : l_2(q - c) - q = 0,$$

$$l_3 = (CAB') \Leftrightarrow h_4 : l_3r - r + c = 0.$$

First suppose that the lines AA', BB', CC' intersect at the point $G = [g_1, g_2]$. Then

$$G \in AA' \Leftrightarrow h_5 : qg_1 - pg_2 = 0,$$

$$G \in BB' \Leftrightarrow h_6 : rg_1 + ag_2 - ar = 0,$$

$$G \in CC' \Leftrightarrow h_7 : cg_1 + sg_2 - sc = 0.$$

We want to prove that (5.3) holds. We derive the condition (5.3) from the given assumptions $h_1, h_2, \ldots, h_7$. In the ideal $I = (h_1, h_2, \ldots, h_7)$ we eliminate all variables up to the independent variables a, c and variables l_1, l_2, l_3, which denote the ratios of points. Elimination of p, q, r, s, g_1, g_2 in the ideal I

```
Use R::=Q[acpqrsg[1..2]l[1..3]];
I:=Ideal(l[1](s-a)-s,l[2]p-p+a,l[2](q-c)-q,l[3]r-r+c,qg[1]-
pg[2],rg[1]+ag[2]-ar,cg[1]+sg[2]-sc);
Elim(p..g[2],I);
```

gives the condition $ac(l_1l_2l_3 + 1) = 0$. Suppose that $a \neq 0$, $c \neq 0$, i.e., $A \neq B$, $A \neq C$ in a triangle ABC and check that $l_1l_2l_3 + 1 = 0$ is the desired condition. The verification

```
Use R::=Q[acpqrsg[1..2]l[1..3]t];
J:=Ideal(l[1](s-a)-s,l[2]p-p+a,l[2](q-c)-q,l[3]r-r+c,qg[1]-
pg[2],rg[1]+ag[2]-ar,act-1,cg[1]+sg[2]-sc);
NF(l[1]l[2]l[3]+1,J);
0
```

confirms that the normal form vanishes and the "only if" part of the statement (5.3) is true.

Similarly we prove the converse implication. Let us assume that for the points A', B', C' the condition (5.3) holds. We want to prove that the straight lines AA', BB', CC' are concurrent. Suppose that G is a common point of AA', BB'. We prove that G lies on CC' as well, i.e., that $cg_1 + sg_2 - sc = 0$ holds. Further suppose, on the basis of a previous implication, that $ac \neq 0$.

```
Use R::=Q[acpqrsg[1..2]l[1..3]t];
K:=Ideal(l[1](s-a)-s,l[2]p-p+a,l[2](q-c)-q,l[3]r-r+c,qg[1]-
pg[2],rg[1]+ag[2]-ar,act-1,l[1]l[2]l[3]+1);
NF(cg[1]+sg[2]-sc,K);
```

We get `NF=0` and the " if " part of the theorem of Ceva is proved.

Remark 5.1.
1) The condition (5.3) is often given in the form

$$\frac{\|AC'\|}{\|C'B\|} \cdot \frac{\|BA'\|}{\|A'C\|} \cdot \frac{\|CB'\|}{\|B'A\|} = 1, \tag{5.5}$$

where $\|AC'\|$ denotes the *signed length* of a segment AC', etc. The signed length $\|XY\|$ of a segment XY is the length $|XY|$, endowed by the sign $+$ or $-$. If we prescribe a direction from X to Y as *positive*, then $\|XY\| = |XY|$ and $\|YX\| = -|XY|$. In any case it holds $\|XY\| = -\|YX\|$. If Z is an inner point of a segment XY, then $\|XZ\|/\|ZY\|$ is positive, if Z lies outside a segment XY, then the ratio $\|XZ\|/\|ZY\|$ is negative (cf. (5.2)).

2) Relations (5.3) and (5.5) are equivalent since

$$(ABC') = -\frac{\|AC'\|}{\|C'B\|}, \ (BCA') = -\frac{\|BA'\|}{\|A'C\|}, \ (CAB') = -\frac{\|CB'\|}{\|B'A\|}.$$

3) The expression of (5.3) in terms of the ratios of points is more suitable for computation on computer than (5.5). Namely, an algebraic translation of the ratio of a point using vectors is simpler than an algebraic translation of the ratio of the length of segments, where square roots of the sum of

squares occur.

4) We proved the theorem of Ceva by computer for the case that the common point G of the lines AA', BB', CC' is a proper point. If the straight lines AA', BB', CC' are pairwise parallel, i.e, they have a common point at infinity, we will proceed in a similar way.

In practice we often encounter a situation when we need to decide if three transversals of a triangle are concurrent. The theorem of Ceva is a tool which offers a quick solution. It suffices to show that for ratios of points $l_1 = (ABC')$, $l_2 = (BCA')$, $l_3 = (CAB')$ the relation (5.3) holds. Let us show a few examples:

a) Medians of a triangle have a common point — the centroid, since the ratios of midpoints of sides equal -1. Then $l_1 = l_2 = l_3 = -1$ and $l_1 l_2 l_3 = -1$.

b) Angle bisectors of a triangle are concurrent at the incenter of a triangle since an angle bisector divides the opposite side of a triangle in the same ratio as is the ratio of adjacent side lengths. Thus $l_1 = -b/a$, $l_2 = -c/b$, $l_3 = -a/c$, where a, b, c are the side lengths of ABC. From here $l_1 l_2 l_3 = -1$.

c) The fact that the altitudes of a triangle are concurrent at the orthocenter can be seen from the theorem of Ceva as well. Suppose that the orthocenter O lies inside ABC (Fig. 5.2). The triangles ACA' and BCB' are similar,

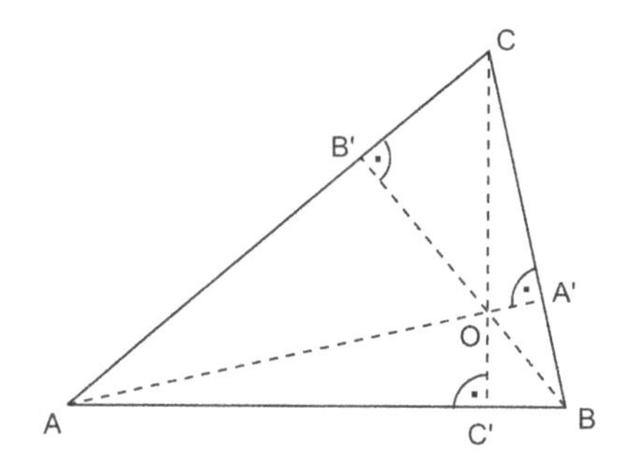

Fig. 5.2 Altitudes of a triangle are concurrent

therefore $\frac{|CB'|}{|A'C|} = \frac{|BB'|}{|AA'|}$. Analogously we get $\frac{|AC'|}{|B'A|} = \frac{|CC'|}{|BB'|}$, $\frac{|BA'|}{|C'B|} = \frac{|AA'|}{|CC'|}$ and from here

$$\frac{\|AC'\|}{\|C'B\|} \cdot \frac{\|BA'\|}{\|A'C\|} \cdot \frac{\|CB'\|}{\|B'A\|} = \frac{|AC'|}{|C'B|} \cdot \frac{|BA'|}{|A'C|} \cdot \frac{|CB'|}{|B'A|}$$

$$= \frac{|CB'|}{|A'C|} \cdot \frac{|AC'|}{|B'A|} \cdot \frac{|BA'|}{|C'B|} = \frac{|BB'|}{|AA'|} \cdot \frac{|CC'|}{|BB'|} \cdot \frac{|AA'|}{|CC'|} = 1.$$

Similarly we proceed if the orthocenter O lies outside ABC.

Now we will show a classical proof of the Ceva's theorem. Instead of (5.3) consider the condition (5.5). We will prove the theorem by means of the so-called *area principle* [49] in which ratios of areas of triangles are used. We distinguish the following two variants of the area principle [51]:

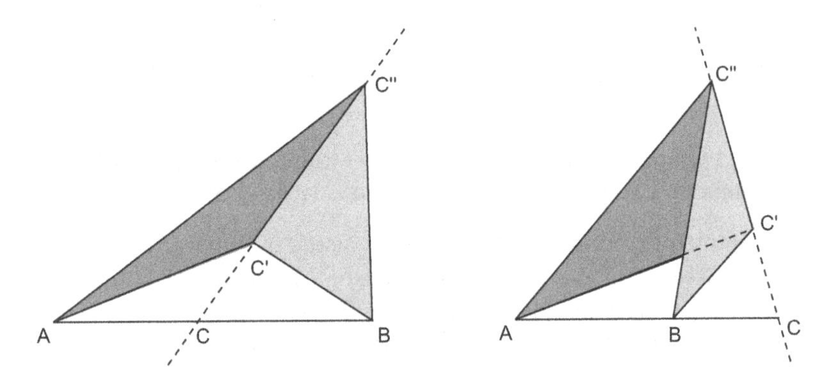

Fig. 5.3 Area principle

APB: The Area Principle for triangles with equal bases: *The areas of two triangles with equal bases are in the same ratio as their heights.*

APH: The Area Principle for triangles with equal heights: *The areas of two triangles with equal heights are in the same ratio as the lengths of their bases.*

The ratio of (unsigned) lengths $|AC|/|CB|$ equals the ratio of areas of triangles $C'C''A$ and $C'C''B$, where C', C'' are two *arbitrary* points on a line passing through the point C (Fig. 5.3). Namely, the ratio of areas of triangles $C'C''A$ and $C'C''B$ equals the ratio of their altitudes on the common base $C'C''$ and it is equal to the ratio of lengths $|AC|/|CB|$.

In this way we can express both the ratio (ABC) and the ratio of signed lengths $\|AC\|/\|CB\|$ and even the ratio of areas. Of course we have to use the *signed* area instead of the (unsigned) area. The signed area of a triangle ABC is positive if we move along the perimeter of a triangle from the vertex A to B and C anti-clockwise. If we move from A to B and C clockwise then the signed area of ABC is negative.

The ratio of signed areas of triangles $C'C''A$ and $C'C''B$ equals the ratio of points (ABC) (see Fig. 5.3 on the left). The area of $C'C''A$ is *positive* since by moving along the perimeter of a triangle from C' to C'' and A we move anti-clockwise. Unlike this, the signed area of a triangle $C'C''B$ is *negative* since moving from C' to C'' and B we move clockwise. Thus the resulting ratio of areas of triangles $C'C''A$, $C'C''B$ is negative which is in accordance with the ratio of points (ABC), since the point C lies between A and B. In Fig. 5.3 on the right both triangles $C'C''A$, $C'C''B$ have positive areas which implies that the ratio of areas of $C'C''A$, $C'C''B$ is positive. It is in accordance with the position of the point C which is outside the segment AB and the ratio (ABC) is positive. We will denote the area of a triangle ABC by $|ABC|$, whereas the signed area by $\|ABC\|$.

Let us return to a classical proof of the theorem of Ceva. First we will assume that the lines AA', BB', CC' intersect at the point G. By the area principle, it holds that (Fig. 5.1):

$$\frac{\|AC'\|}{\|C'B\|} = \frac{\|CAG\|}{\|BCG\|},$$

where $\|CAG\|$ is the signed area of a triangle CAG, etc. Then we can easily check that

$$\frac{\|BA'\|}{\|A'C\|} = \frac{\|ABG\|}{\|CAG\|} \quad \text{and} \quad \frac{\|CB'\|}{\|B'A\|} = \frac{\|BCG\|}{\|ABG\|}.$$

Hence

$$\frac{\|AC'\|}{\|C'B\|} \cdot \frac{\|BA'\|}{\|A'C\|} \cdot \frac{\|CB'\|}{\|B'A\|} = \frac{\|CAG\|}{\|BCG\|} \cdot \frac{\|ABG\|}{\|CAG\|} \cdot \frac{\|BCG\|}{\|ABG\|} = 1.$$

We showed that (5.5) is a necessary condition for the lines AA', BB', CC' being concurrent.

Now we will show that the condition (5.5) is sufficient, i.e., the validity of (5.5) implies that the lines AA', BB', CC' are concurrent (Fig. 5.1). We will prove by contradiction. Suppose that (5.5) holds and that the lines AA', BB', CC' do not intersect at a common point. Let G be a common point of AA', BB'. Through G we construct the line CG which intersects AB at the point $C'' \neq C'$. By the first part of the Ceva's theorem applied on the lines AA', BB', CC''

$$\frac{\|AC''\|}{\|C''B\|} \cdot \frac{\|BA'\|}{\|A'C\|} \cdot \frac{\|CB'\|}{\|B'A\|} = 1$$

holds. A comparison with the assumption (5.5) gives

$$\frac{\|AC''\|}{\|C''B\|} = \frac{\|AC'\|}{\|C'B\|}.$$

From here $C' = C''$ follows, which is a contradiction.

Now we will suppose that the lines AA', BB', CC' are parallel (Fig. 5.4). A parallel projection of the points A', C, B' onto the line AB in the direc-

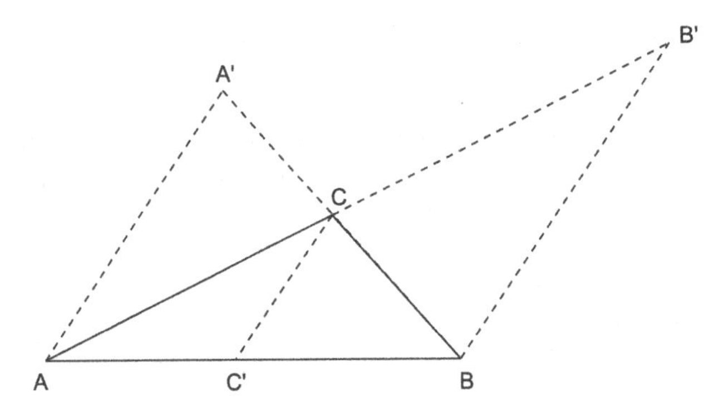

Fig. 5.4 Ceva's theorem: If AA', BB', CC' are parallel then $\frac{\|AC'\|}{\|C'B\|} \cdot \frac{\|BA'\|}{\|A'C\|} \cdot \frac{\|CB'\|}{\|B'A\|} = 1$

tion AA' gives the points A, C', B respectively. Knowing that a parallel projection preserves ratios of points (or ratios of signed lengths) we can write

$$\frac{\|AC'\|}{\|C'B\|} \cdot \frac{\|BA'\|}{\|A'C\|} \cdot \frac{\|CB'\|}{\|B'A\|} = \frac{\|AC'\|}{\|C'B\|} \cdot \frac{\|BA\|}{\|AC'\|} \cdot \frac{\|C'B\|}{\|BA\|} = 1. \tag{5.6}$$

This proves one part of the Ceva's theorem when AA', BB', CC' are parallel.

The proof of the converse implication is similar to the corresponding proof above when AA', BB', CC' are concurrent.

5.1.1 *Generalization of the theorem of Ceva*

In this part we will deal with a generalization of the Ceva's theorem on a quadrilateral and pentagon in a plane [49]. First we will focus on a pentagon.

Let $ABCDE$ be a planar pentagon and M an arbitrary point in the plane of a pentagon. Denote by M_1, M_2, M_3, M_4, M_5 intersections of the lines AM, BM, CM, DM, EM with the sides CD, DE, EA, AB, BC (extended if necessary) respectively. What is the relation between ratios of points (CDM_1), (DEM_2), (EAM_3), (ABM_4), (BCM_5)?

We will derive a missing relation from the given assumptions.

Choose an affine coordinate system so that $A = [0,0]$, $B = [1,0]$, $C = [x,y]$, $D = [u,v]$, $E = [0,1]$, $M_1 = [k_1,k_2]$, $M_2 = [l_1,l_2]$, $M_3 = [0,m_2]$, $M_4 = [n_1,0]$, $M_5 = [o_1,o_2]$, $M = [p,q]$ (Fig. 5.5).

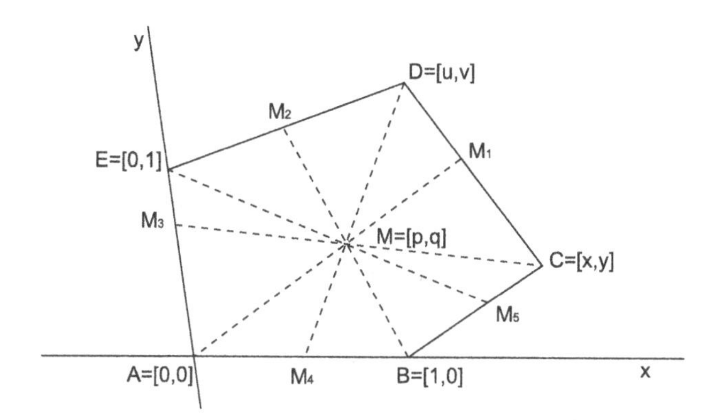

Fig. 5.5 Ceva's theorem for a pentagon — computer proof

Denote the ratios of points which are of our concern by $d_1 = (CDM_1)$, $d_2 = (DEM_2)$, $d_3 = (EAM_3)$, $d_4 = (ABM_4)$, $d_5 = (BCM_5)$. Then the following relations hold:

$M_1 \in CD \Leftrightarrow h_1 : k_1y + xv + k_2u - yu - k_1v - k_2x = 0,$

$M_2 \in DE \Leftrightarrow h_2 : l_1v + u - l_1 - l_2u = 0,$

$M_5 \in BC \Leftrightarrow h_3 : y + o_2x - o_1y - o_2 = 0,$

$d_4 = (ABM_4) \Leftrightarrow h_4 : d_4(n_1 - 1) - n_1 = 0,$

$d_5 = (BCM_5) \Leftrightarrow h_5 : d_5(o_1 - x) - o_1 + 1 = 0,$

$h_6 : d_5(o_2 - y) - o_2 = 0,$

$d_1 = (CDM_1) \Leftrightarrow h_7 : d_1(k_1 - u) - k_1 + x = 0,$

$h_8 : d_1(k_2 - v) - k_2 + y = 0,$

$d_2 = (DEM_2) \Leftrightarrow h_9 : d_2l_1 - l_1 + u = 0,$

$h_{10} : d_2(l_2 - 1) - l_2 + v = 0,$

$d_3 = (EAM_3) \Leftrightarrow h_{11} : d_3m_2 - m_2 + 1 = 0,$

$M \in AM_1 \Leftrightarrow h_{12} : k_1q - k_2p = 0,$

$$M \in BM_2 \Leftrightarrow h_{13} : l_2 + l_1 q - l_2 p - q = 0,$$

$$M \in CM_3 \Leftrightarrow h_{14} : py + m_2 x - m_2 p - qx = 0,$$

$$M \in DM_4 \Leftrightarrow h_{15} : pv + n_1 q - n_1 v - uq = 0,$$

$$M \in EM_5 \Leftrightarrow h_{16} : p + o_1 q - o_1 - o_2 p = 0.$$

In the ideal $(h_1, h_2, \ldots, h_{16})$ we successively eliminate all variables except $d_1, d_2, \ldots, d_5$. First we eliminate variables k_1, k_2, l_1, l_1 from all polynomials which contain these variables. Then we add polynomials which contain variables $m_1, m_2, n_1, n_2, o_1, o_2$ in the acquired elimination ideal and eliminate these variables. In the end we eliminate the remaining variables x, y, u, v, p, q. The elimination leads to the desired condition holding among respective ratios of points:

$$(CDM_1) \cdot (DEM_2) \cdot (EAM_3) \cdot (ABM_4) \cdot (BCM_5) = -1. \tag{5.7}$$

By verifying whether the resulting condition (5.7) obeys the statement, we have to rule out the case $p + q = 1$ when the points B, E, M are collinear, that is, $M_2 = E$, and the ratio DEM_2 is not defined. Entering

```
Use R::=Q[xyuvpqk[1..2]l[1..2]m[1..2]n[1..2]o[1..2]d[1..5]tr];
I:=Ideal(k[1]y+xv+k[2]u-yu-k[1]v-k[2]x,l[1]v+u-l[1]-l[2]u,
d[1](k[1]-u)-k[1]+x,d[1](k[2]-v)-k[2]+y,l[2]+l[1]q-l[2]p-q,
d[2]l[1]-l[1]+u,d[2](l[2]-1)-l[2]+v,k[1]q-k[2]p,y+o[2]x-o[1]y
-o[2],d[4](n[1]-1)-n[1],d[5](o[1]-x)-o[1]+1,d[5](o[2]-y)-o[2],
d[3]m[2]-m[2]+1,py+m[2]x-m[2]p-qx,pv+n[1]q-n[1]v-uq,p+o[1]q-
o[1]-o[2]p,(p+q-1)r-1,(d[1]d[2]d[3]d[4]d[5]+1)t-1);
NF(1,I);
```

CoCoA returns `NF(1,I)=0`. We get the following generalization of the Ceva's theorem for a pentagon [49]:

Let $ABCDE$ be a planar pentagon and M be an arbitrary point of the plane of a pentagon. Let M_1, M_2, M_3, M_4, M_5 be the intersections of the lines AM, BM, CM, DM, EM with the sides CD, DE, EA, AB, BC (extended if necessary) respectively. Then

$$(ABM_4) \cdot (BCM_5) \cdot (CDM_1) \cdot (DEM_2) \cdot (EAM_3) = -1. \tag{5.8}$$

Remark 5.2. The condition (5.8) is equivalent to the condition

$$\frac{\|AM_4\|}{\|M_4B\|} \cdot \frac{\|BM_5\|}{\|M_5C\|} \cdot \frac{\|CM_1\|}{\|M_1D\|} \cdot \frac{\|DM_2\|}{\|M_2E\|} \cdot \frac{\|EM_3\|}{\|M_3A\|} = 1. \tag{5.9}$$

A classical proof of the Ceva's theorem for a pentagon is based on the area principle.

Assume that M_1, M_2, M_3, M_4, M_5 are intersections of the lines AM, BM, CM, DM, EM with the sides CD, DE, EA, AB, BC of a pentagon $ABCDE$ respectively (Fig. 5.6). It holds that

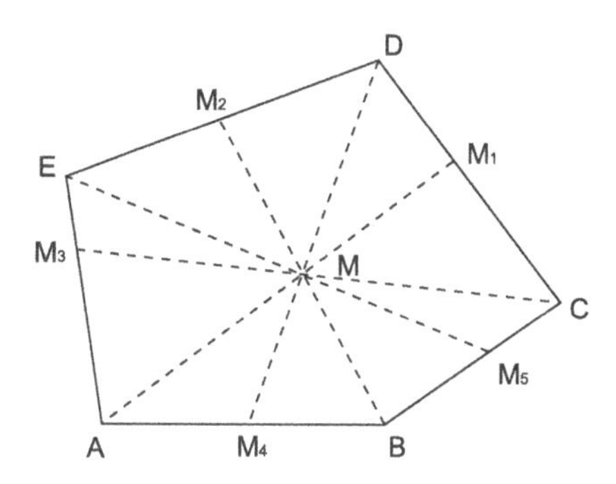

Fig. 5.6 Ceva's theorem for a pentagon — classical proof

$$\frac{\|AM_4\|}{\|M_4B\|} = \frac{\|DAM\|}{\|BDM\|}, \frac{\|BM_5\|}{\|M_5C\|} = \frac{\|EBM\|}{\|CEM\|}, \frac{\|CM_1\|}{\|M_1D\|} = \frac{\|ACM\|}{\|DAM\|},$$
$$\frac{\|DM_2\|}{\|M_2E\|} = \frac{\|BDM\|}{\|EBM\|}, \frac{\|EM_3\|}{\|M_3A\|} = \frac{\|CEM\|}{\|ACM\|}.$$

Thus we get

$$\frac{\|AM_4\|}{\|M_4B\|} \cdot \frac{\|BM_5\|}{\|M_5C\|} \cdot \frac{\|CM_1\|}{\|M_1D\|} \cdot \frac{\|DM_2\|}{\|M_2E\|} \cdot \frac{\|EM_3\|}{\|M_3A\|} =$$
$$\frac{\|DAM\|}{\|BDM\|} \cdot \frac{\|EBM\|}{\|CEM\|} \cdot \frac{\|ACM\|}{\|DAM\|} \cdot \frac{\|BDM\|}{\|EBM\|} \cdot \frac{\|CEM\|}{\|ACM\|} = 1$$

which is the condition (5.9). The theorem is proved.

The theorem we have just proved is a special case of a theorem which was published by B. Grünbaum and G. C. Shephard [49] in 1995. It is as follows:

Theorem 5.2 (Ceva's theorem for n-gons). *Let $A_1A_2 \ldots A_n$ be an arbitrary n-gon, and M a given point. Let M_i be the intersection of the line MA_i with the side $A_{i-k}A_{i+k}$ or its extension, where $i = 1, 2, \ldots, n$ and $1 \leq k < n/2$. Then*

$$\prod_{i=1}^{n} \frac{\|A_{i-k}M_i\|}{\|M_iA_{i+k}\|} = 1. \tag{5.10}$$

For the values $n = 5$ and $k = 2$ we get (5.9) from the Ceva's theorem for a pentagon.

Let us show the Ceva's theorem for the case of a quadrilateral $A_1A_2A_3A_4$, i.e., for $n = 4$ and $k = 1$ (Fig. 5.7). By (5.10) it holds that

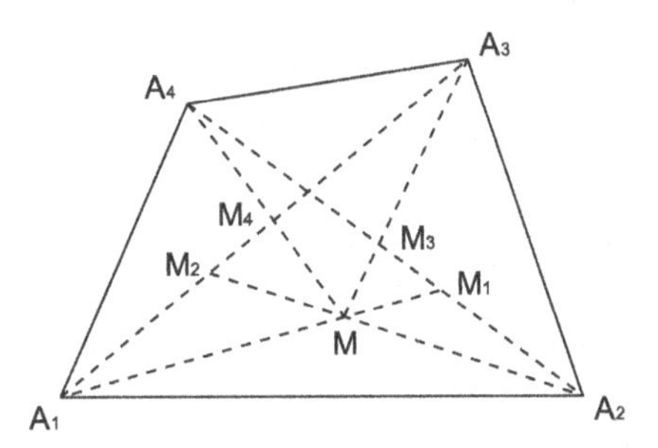

Fig. 5.7 Ceva's theorem for a quadrilateral

$$\frac{\|A_4M_1\|}{\|M_1A_2\|} \cdot \frac{\|A_1M_2\|}{\|M_2A_3\|} \cdot \frac{\|A_2M_3\|}{\|M_3A_4\|} \cdot \frac{\|A_3M_4\|}{\|M_4A_1\|} = 1. \tag{5.11}$$

The relation (5.11) is easy to check both by computer and classically. We leave it for the reader to solve.

It is obvious that the converse theorem to the given generalization of Ceva's theorem for a pentagon, unlike the Ceva's theorem for a triangle, does not hold. We will formulate an analogue to this case in a rather weaker form (see [123]):

If four transversals AM_1, BM_2, CM_3, DM_4 of a pentagon $ABCDE$ are passing through a point M and the condition (5.7) holds, then the transversal EM_5 passes through M as well.

A proof by computer is similar to the proof of Ceva's theorem for a pentagon. With the same notation it suffices to show that the polynomial $h_{16} : p + o_1q - o_1 - o_2p$ is an element of the ideal $J = (h_1, h_2, \ldots, h_{15})$. Assuming that $y \neq 0$, i.e., the vertices A, B, C are not collinear, we get `NF(h16,J)=0`.

5.2 Menelaus' theorem

The theorem of Menelaus (Menelaus, 1st century A.D.) is a well-known theorem of elementary geometry in the plane. It reads:

Theorem 5.3 (Menelaus). *Let A', B', C' be three points on the sides BC, CA, AB or their extensions of a triangle ABC respectively. Then the points A', B', C' are collinear if and only if*

$$(ABC') \cdot (BCA') \cdot (CAB') = 1. \tag{5.12}$$

To prove the theorem choose an affine coordinate system so that $A = [0, 0]$, $B = [a, 0]$, $C = [0, b]$, $A' = [p, q]$, $B' = [0, r]$, $C' = [s, 0]$ (Fig. 5.8). Denote

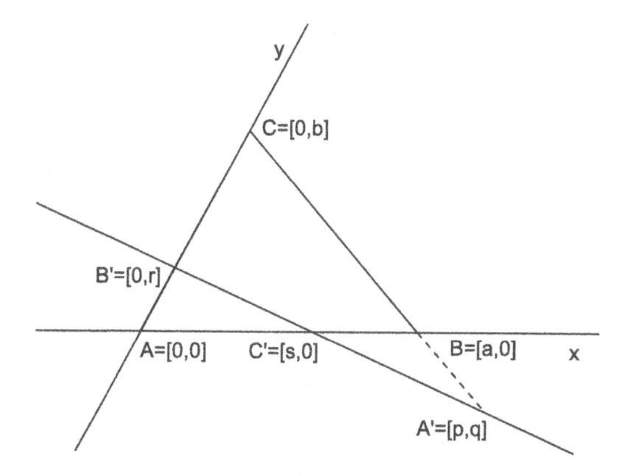

Fig. 5.8 Menelaus' theorem

by $l_1 = (ABC')$ the ratio of a point C' with respect to the points A, B and analogously $l_2 = (BCA')$ and $l_3 = (CAB')$. Then

$l_1 = (ABC') \Leftrightarrow h_1 : l_1(s - a) - s = 0,$

$l_2 = (BCA') \Leftrightarrow h_2 : l_2 p - p + a = 0,$

$h_3 : l_2(q - b) - q = 0,$

$l_3 = (CAB') \Leftrightarrow h_4 : l_3 r - r + b = 0.$

We will show that (5.12) is a necessary condition for the points A', B', C' to be collinear.

A', B', C' are collinear $\Leftrightarrow h_5 : pr + qs - rs = 0.$

The conclusion c has the form

$c : l_1 l_2 l_3 - 1 = 0.$

In addition we will assume that $ab \neq 0$. The normal form of the conclusion polynomial c with respect to the Gröbner basis of the ideal $I = (h_1, h_2, h_3, h_4, h_5, abt - 1)$, where t is a slack variable,

```
Use R ::=Q[abrspql[1..3]t];
I:=Ideal(l[1](s-a)-s,l[2]p-p+a,l[2](q-b)-q,l[3]r-r+b,pr+qs-rs,
abt-1);
NF(l[1]l[2]l[3]-1,I);
```

equals zero, hence (5.12) is a necessary condition.
Now we prove the sufficiency of (5.12). We are to show that from (5.12), collinearity of A', B', C' follows. Computing the normal form of h_5 with respect to the ideal $J = (h_1, h_2, h_3, h_4, abt - 1, c)$ we enter

```
Use R ::= Q[abrspql[1..3]t];
J:=Ideal(l[1](s-a)-s,l[2]p-p+a,l[2](q-b)-q,l[3]r-r+b,abt-1,
l[1]l[2]l[3]-1);
NF(pr+qs-rs,J);
```

and get the answer 0. Menelaus' theorem is proved.

Remark 5.3. The condition (5.12) is often given in the form

$$\frac{\|AC'\|}{\|C'B\|} \cdot \frac{\|BA'\|}{\|A'C\|} \cdot \frac{\|CB'\|}{\|B'A\|} = -1, \tag{5.13}$$

where $\|XY\|$ denotes the *signed length* of a segment XY.

Let us show a classical proof of the Menelaus' theorem. It is based on the area principle (Fig. 5.9).

Suppose that a straight line p intersects the sides AB, BC, CA of a triangle at the points C', A', B' respectively. We will show that (5.13) holds. We express the ratios from (5.13) in terms of the ratios of the areas of triangles $B'C'A$, $B'C'B$ and $B'C'C$. Note that all three triangles have a common side $B'C'$. Then

$$\frac{\|AC'\|}{\|C'B\|} = -\frac{\|B'C'A\|}{\|B'C'B\|},$$

where $\|B'C'A\|$ denotes the signed area of $B'C'A$, etc. Similarly,

$$\frac{\|BA'\|}{\|A'C\|} = -\frac{\|B'C'B\|}{\|B'C'C\|} \quad \text{and} \quad \frac{\|CB'\|}{\|B'A\|} = -\frac{\|B'C'C\|}{\|B'C'A\|}.$$

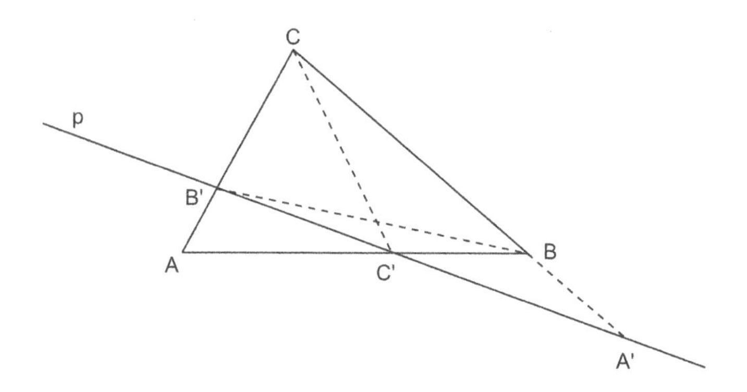

Fig. 5.9 Proof of Menelaus' theorem by the area principle

Whence

$$\frac{\|AC'\|}{\|C'B\|} \cdot \frac{\|BA'\|}{\|A'C\|} \cdot \frac{\|CB'\|}{\|B'A\|} = -\frac{\|B'C'A\|}{\|B'C'B\|} \cdot \frac{\|B'C'B\|}{\|B'C'C\|} \cdot \frac{\|B'C'C\|}{\|B'C'A\|} = -1.$$

Conversely we can prove, similarly as in the Ceva's theorem, that the condition (5.13) implies the collinearity of the points A', B', C'.

Remark 5.4.

1) Notice that if we use in (5.5) and (5.13) unsigned lengths instead of signed ones, then the necessary conditions for both theorems — Ceva's and Menelaus' — are the same, namely

$$\frac{|AC'|}{|C'B|} \cdot \frac{|BA'|}{|A'C|} \cdot \frac{|CB'|}{|B'A|} = 1. \tag{5.14}$$

(5.14) is a necessary and sufficient condition for the lines AA', BB', CC' being concurrent *or* for points A', B', C' being collinear. To distinguish both theorems we need an expression finer than (5.14). That is why we used the ratio of points or signed lengths. Whereas the ratio of points determines the position of a point on a line uniquely, the ratio of lengths is not unique. The reason is that $|(ABC')| = \frac{|AC'|}{|C'B|}$.

2) The Ceva's and Menelaus' theorems make it possible to prove, for instance, theorems of Pappus, Desargues or Pascal, which are basic theorems of projective geometry (see [5]).

5.2.1 *Generalization of Menelaus' theorem*

Generalization of the Menelaus' theorem will be carried out in two ways. First we show its generalization on polygons in the plane. In the second part we will generalize the Menelaus' theorem on skew polygons in the three dimensional space.

We will solve the following problem:

Given a quadrilateral $ABCD$ in the plane and points K, L, M, N on its sides AB, BC, CD, DA or their extensions respectively. Suppose that the points K, L, M, N are collinear. Find a condition which is fulfilled by the ratios of points (ABK), (BCL), (CDM), (DAN).

Denote $p_1 = (ABK)$, $p_2 = (BCL)$, $p_3 = (CDM)$, $p_4 = (DAN)$ and introduce an affine coordinate system so that $A = [0, 0]$, $B = [a, 0]$, $C = [b, c]$, $D = [0, d]$, $K = [k_1, 0]$, $L = [l_1, l_2]$, $M = [m_1, m_2]$, $N = [0, n_2]$ (Fig. 5.10). The situation may be described in the following way:

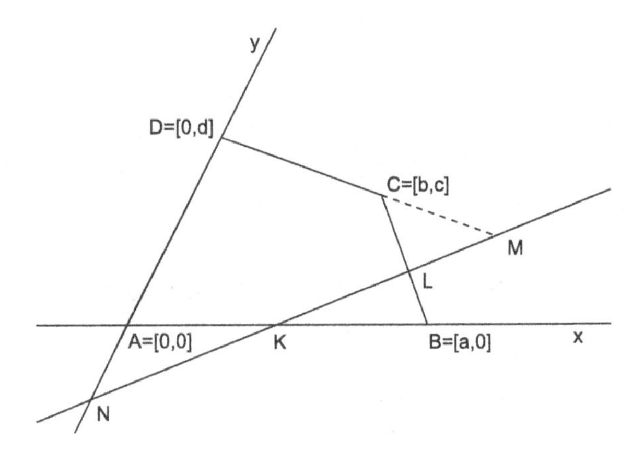

Fig. 5.10 Menelaus' theorem for a quadrilateral — computer proof

$$L \in BC \Leftrightarrow h_1 : l_1c + l_2a - ac - l_2b = 0,$$

$$M \in CD \Leftrightarrow h_2 : m_1c + bd - m_1d - m_2b = 0,$$

$$p_1 = (ABK) \Leftrightarrow h_3 : p_1(k_1 - a) - k_1 = 0,$$

$$p_2 = (BCL) \Leftrightarrow h_4 : p_2(l_1 - b) - l_1 + a = 0,$$

$$h_5 : p_2(l_2 - c) - l_2 = 0,$$

$$p_3 = (CDM) \Leftrightarrow h_6 : p_3m_1 - m_1 + b = 0,$$

$$h_7 : p_3(m_2 - d) - m_2 + c = 0,$$

$$p_4 = (DAN) \Leftrightarrow h_8 : p_4n_2 - n_2 + d = 0,$$

K, L, M, N are collinear $\Leftrightarrow h_9 : k_1l_2 + l_1m_2 - l_2m_1 - k_1m_2 = 0,$

$$h_{10} : k_1l_2 + l_1n_2 - k_1n_2 = 0.$$

Elimination of all variables besides $a, b, c, d, p_1, p_2, p_3, p_4$ in the ideal $(h_1, h_2, \ldots, h_{10})$ yields the condition $ac(p_1p_2p_3p_4 - 1) = 0$. Suppose that $ac \neq 0$, i.e., $A \neq B$ and A, B, C are not collinear. Then

```
Use R::=Q[abcdk[1..2]l[1..2]m[1..2]n[1..2]p[1..4]t];
I:=Ideal(l[1]c+l[2]a-ac-l[2]b,m[1]c+bd-m[1]d-m[2]b,p[1](k[1]-
a)-k[1],p[2](l[1]-b)-l[1]+a,p[2](l[2]-c)-l[2],p[3]m[1]-m[1]+b,
p[3](m[2]-d)-m[2]+c,p[4]n[2]-n[2]+d,k[1]l[2]+l[1]m[2]-l[2]m[1]
-k[1]m[2],k[1]l[2]+l[1]n[2]-k[1]n[2],at-1);
NF(p[1]p[2]p[3]p[4]-1,I);
0
```

the normal form of the polynomial $p_1p_2p_3p_4 - 1$ with respect to the Gröbner basis of the ideal I equals zero. It means that the following generalization of the Menelaus' theorem for a quadrilateral in the plane holds [49]:

Given a quadrilateral $ABCD$ in the plane and points K, L, M, N on its sides AB, BC, CD, DA or their extensions respectively. Suppose that the points K, L, M, N are collinear. Then

$$(ABK) \cdot (BCL) \cdot (CDM) \cdot (DAN) = 1. \tag{5.15}$$

Now we will prove the equality (5.15) in a classical way by means of the area principle. This also indicates the method of how to prove the analogue of the Menelaus' theorem for an arbitrary n-gon. Using this we should realize the "power" of the area principle.
Assume that a line p intersects the sides AB, BC, CD, DA of a quadrilateral at the points K, L, M, N respectively (Fig. 5.11). We will prove that

$$\frac{\|AK\|}{\|KB\|} \cdot \frac{\|BL\|}{\|LC\|} \cdot \frac{\|CM\|}{\|MD\|} \cdot \frac{\|DN\|}{\|NA\|} = 1 \tag{5.16}$$

holds.

We have relations

$$\frac{\|AK\|}{\|KB\|} = -\frac{\|KLA\|}{\|KLB\|}, \qquad \frac{\|BL\|}{\|LC\|} = -\frac{\|KLB\|}{\|KLC\|},$$

$$\frac{\|CM\|}{\|MD\|} = -\frac{\|KLC\|}{\|KLD\|}, \qquad \frac{\|DN\|}{\|NA\|} = -\frac{\|KLD\|}{\|KLA\|}.$$

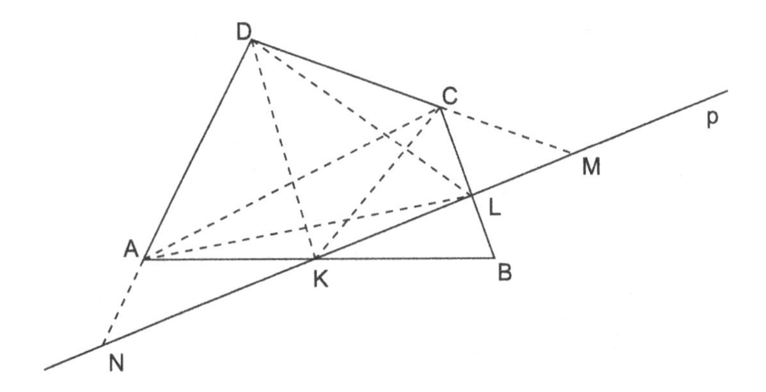

Fig. 5.11 Menelaus' theorem for a quadrilateral — classical proof by area principle

From here we get

$$\frac{\|AK\|}{\|KB\|} \cdot \frac{\|BL\|}{\|LC\|} \cdot \frac{\|CM\|}{\|MD\|} \cdot \frac{\|DN\|}{\|NA\|} =$$

$$\frac{\|KLA\|}{\|KLB\|} \cdot \frac{\|KLB\|}{\|KLC\|} \cdot \frac{\|KLC\|}{\|KLD\|} \cdot \frac{\|KLD\|}{\|KLA\|} = 1$$

and the Menelaus's theorem for a quadrilateral is proved.

The relation (5.15) is a special case of the theorem which was published by B. Grünbaum and G. C. Shephard in 1995 (see [49]). It is as follows:

Theorem 5.4 (Menelaus theorem for n-gons). *Let $A_1A_2 \ldots A_n$ be an arbitrary n-gon and suppose that a transversal cuts the line A_iA_{i+1} at M_i, where $i = 1, 2, \ldots, n$. Then*

$$\prod_{i=1}^{n} \frac{\|A_iM_i\|}{\|M_iA_{i+1}\|} = (-1)^n. \tag{5.17}$$

Hence, for polygons with an even number of vertices we obtain in (5.17) the result 1, whereas the polygons with an odd number of vertices give -1. Let us see the case of a pentagon.

Let the notation be as above and consider a pentagon $A_1A_2A_3A_4A_5$. As the automatic proof is similar to that for the case $n = 4$, we omit it and show only a classical proof.

This proof is very simple and is worth mentioning. The proof is based on the fact that a parallel projection preserves ratios of points.

Project the vertices A_1, A_2, A_3, A_4, A_5 of a pentagon and the points M_1, M_2, M_3, M_4, M_5 in the direction given by the straight line p onto a straight line q (Fig. 5.12). The vertices A_1, A_2, A_3, A_4, A_5 project onto

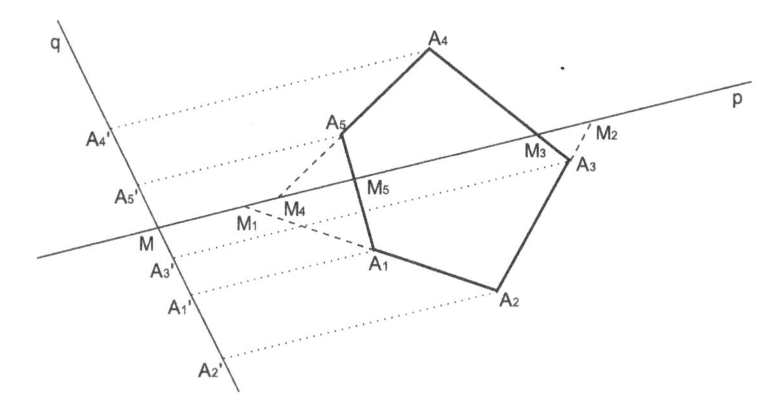

Fig. 5.12 Menelaus' theorem for a pentagon — classical proof using parallel projection

$A_1', A_2', A_3', A_4', A_5'$ respectively and the points M_1, M_2, M_3, M_4, M_5 project onto a single point M. Then

$$\frac{\|A_1M_1\|}{\|M_1A_2\|} \cdot \frac{\|A_2M_2\|}{\|M_2A_3\|} \cdot \frac{\|A_3M_3\|}{\|M_3A_4\|} \cdot \frac{\|A_4M_4\|}{\|M_4A_5\|} \cdot \frac{\|A_5M_5\|}{\|M_5A_1\|} =$$

$$\frac{\|A_1'M\|}{\|MA_2'\|} \cdot \frac{\|A_2'M\|}{\|MA_3'\|} \cdot \frac{\|A_3'M\|}{\|MA_4'\|} \cdot \frac{\|A_4'M\|}{\|MA_5'\|} \cdot \frac{\|A_5'M\|}{\|MA_1'\|} = -1$$

since we have an odd number of pairs $\|A_i'M\|, \|MA_i'\|$ with the ratio -1.

5.2.2 *Generalization of Menelaus' theorem in space*

In the next part we will generalize the theorem of Menelaus into the three dimensional space. Instead of plane polygons now we consider skew polygons which are cut by an arbitrary plane. We search for a condition, similarly as in the plane, holding for the intersections of the plane and a polygon.

We will demonstrate this generalization on a skew quadrilateral. We search for a condition such that four points on the sides of a skew quadrilateral $ABCD$ are complanar [99].

Suppose we are given a skew quadrilateral $ABCD$ and on its sides

AB, BC, CD, DA or their extensions points K, L, M, N respectively. Find a necessary and sufficient condition for the ratios of points $(ABK), (BCL), (CDM), (DAN)$ such that the points K, L, M, N are complanar.

Let us choose an affine coordinate system so that $A = [0,0,0]$, $B = [b,0,0]$, $C = [0,0,c]$, $D = [0,d,0]$, $K = [k_1,0,0]$, $L = [l_1,0,l_3]$, $M = [0,m_2,m_3]$, $N = [0,n_2,0]$ (Fig. 5.13). Denote the ratios of points by $p_1 = (ABK)$, $p_2 = (BCL)$, $p_3 = (CDM)$, $p_4 = (DAN)$. The following relations hold:

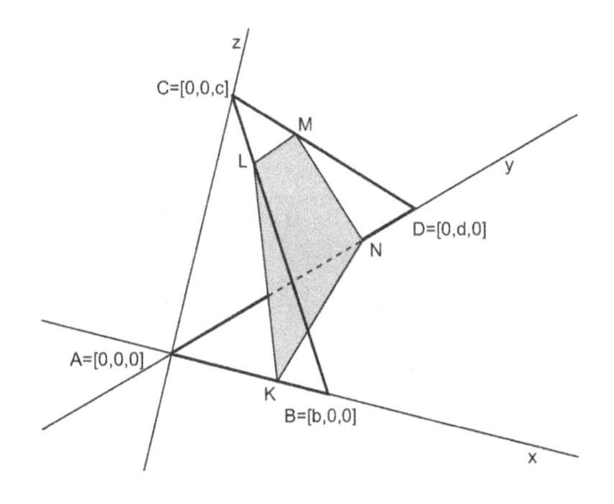

Fig. 5.13 Menelaus' theorem in space

$L \in BC \Leftrightarrow h_1 : c(l_1 - b) + bl_3 = 0,$

$M \in CD \Leftrightarrow h_2 : c(m_2 - d) + dm_3 = 0,$

$p_1 = (ABK) \Leftrightarrow h_3 : p_1(k_1 - b) - k_1 = 0,$

$p_2 = (BCL) \Leftrightarrow h_4 : p_2 l_1 - l_1 + b = 0,$

$h_5 : p_2(l_3 - c) - l_3 = 0,$

$p_3 = (CDM) \Leftrightarrow h_6 : p_3 m_3 - m_3 + c = 0$

$h_7 : p_3(m_2 - d) - m_2 = 0,$

$p_4 = (DAN) \Leftrightarrow h_8 : p_4 n_2 - n_2 + d = 0,$

K, L, M, N are complanar $\Leftrightarrow h_9 : k_1 l_3 n_2 - k_1 m_3 n_2 - k_1 l_3 m_2 + l_1 m_3 n_2 = 0.$

We search for a condition such that the points K, L, M, N are complanar.

To do this we add the conclusion h_9 to the ideal $I = (h_1, h_2, \ldots, h_8)$ and eliminate variables $k_1, l_1, l_3, m_2, m_3, n_2$ in the ideal $J = I \cup \{h_9\}$. We get

```
Use R::=Q[bcdk[1..3]l[1..3]m[1..3]n[1..3]p[1..4]];
J:=Ideal(c(l[1]-b)+bl[3],c(m[2]-d)+dm[3],p[1](k[1]-b)-k[1],
p[2]l[1]-l[1]+b,p[2](l[3]-c)-l[3],p[3]m[3]-m[3]+c,p[3](m[2]
-d)-m[2],p[4]n[2]-n[2]+d,k[1]l[3]n[2]-k[1]m[3]n[2]-
k[1]l[3]m[2]+l[1]m[3]n[2]);
Elim(k[1]..n[3],J);
```

the equation $bcd(p_1p_2p_3p_4 - 1) = 0$. Assuming that $bcd \neq 0$ we get the only condition

$$p_1p_2p_3p_4 = 1. \tag{5.18}$$

We proved that the equality (5.18) is a necessary condition for the points K, L, M, N to be complanar.

Now we verify if (5.18) is also a sufficient condition, i.e., we are to show that h_9 belongs to the ideal $K = I \cup \{bcdt - 1\} \cup \{p_1p_2p_3p_4 - 1\}$, where $bcdt - 1 = 0$ is a non-degeneracy condition. We obtain

```
Use R::=Q[bcdk[1..3]l[1..3]m[1..3]n[1..3]p[1..4]t];
K:=Ideal(c(l[1]-b)+bl[3],c(m[2]-d)+dm[3],p[1](k[1]-b)-k[1],
p[2]l[1]-l[1]+b,p[2](l[3]-c)-l[3],p[3]m[3]-m[3]+c,p[3](m[2]
-d)-m[2],p[4]n[2]-n[2]+d,bcdt-1,p[1]p[2]p[3]p[4]- 1);
NF(k[1]l[3]n[2]-k[1]m[3]n[2]- k[1]l[3]m[2]+l[1]m[3]n[2],K);
0
```

the answer 0, i.e., the condition (5.18) is sufficient as well. We proved a statement [99]:

Suppose we are given a skew quadrilateral $ABCD$ and on its sides AB, BC, CD, DA or their extensions points K, L, M, N respectively. The points K, L, M, N are complanar if and only if

$$(ABK) \cdot (BCL) \cdot (CDM) \cdot (DAN) = 1. \tag{5.19}$$

Let us prove the statement above classically (see [99]). We will prove the "only if" part. Consider a straight line p which is perpendicular to the plane K, L, M, N and project a skew quadrilateral A, B, C, D orthogonally onto the straight line p. Then the points K, L, M, N will project into one point, say P, and the points A, B, C, D into the points A', B', C', D'. Taking into account that a parallel projection preserves the ratios of points we get

$$\frac{\|AK\|}{\|KB\|}\cdot\frac{\|BL\|}{\|LC\|}\cdot\frac{\|CM\|}{\|MD\|}\cdot\frac{\|DN\|}{\|NA\|}=$$

$$\frac{\|A'P\|}{\|PB'\|}\cdot\frac{\|B'P\|}{\|PC'\|}\cdot\frac{\|C'P\|}{\|PD'\|}\cdot\frac{\|D'P\|}{\|PA'\|}=1.$$

The "only if" part of the statement is proved.

To conclude the part related to the Menelaus' theorem we will give the Menelaus' theorem for an arbitrary skew n-gon. It is as follows [99]:

Theorem 5.5 (Menelaus' theorem for skew n-gons).
Let $A_1A_2\ldots A_n$ be an arbitrary skew n-gon and suppose that a plane cuts the line A_iA_{i+1} at M_i, where $i=1,2,\ldots,n$. Then

$$\prod_{i=1}^{n}\frac{\|A_iM_i\|}{\|M_iA_{i+1}\|}=(-1)^n. \tag{5.20}$$

Remark 5.5.
1) Proving the previous theorem by computer we could take affine coordinates of the vertices of a skew quadrilateral $ABCD$ also in this way $A=[0,0,0]$, $B=[1,0,0]$, $C=[0,1,0]$, $D=[0,0,1]$, instead of $A=[0,0,0]$, $B=[b,0,0]$, $C=[0,c,0]$, $D=[0,0,d]$ to save *three* variables b,c,d.

2) Expressions of the type $p_1p_2p_3p_4$ which occur in Ceva's and Menelaus' theorems are called *cyclic products* (see [123]).

5.3 Theorem of Euler

The following theorem of Euler is a generalization of the well-known property of medians in a triangle:

If AA', BB', CC' are the medians in a triangle ABC then $|GA'|/|AA'|=|GB'|/|BB'|=|GC'|/|CC'|=1/3$, that is

$$\frac{|GA'|}{|AA'|}+\frac{|GB'|}{|BB'|}+\frac{|GC'|}{|CC'|}=1 \tag{5.21}$$

(Fig. 5.14).

It is interesting that the equality (5.21) holds not only for the centroid but for an arbitrary *inner* point G of a triangle. In addition, if we use in (5.21) the signed lengths then the theorem is valid even for *all* points of the plane of a triangle for which (5.21) is defined.

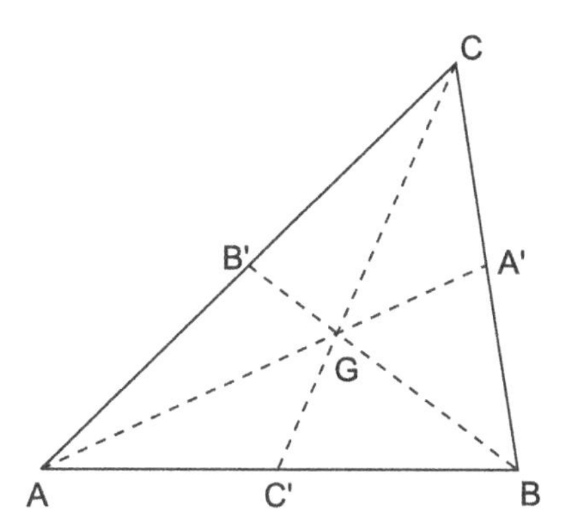

Fig. 5.14 If G is the centroid of a triangle ABC, then $\frac{|GA'|}{|AA'|} + \frac{|GB'|}{|BB'|} + \frac{|GC'|}{|CC'|} = 1$

We will prove the theorem which was published by L. Euler in 1812 [36, 48], but was in fact accepted for publication in 1780, as G. C. Shephard says [122] (cf. [10]).

The theorem of Euler in its generalized form reads:

Theorem 5.6. *Let G be an arbitrary point in the plane of a triangle ABC and let the lines AG, BG, CG intersect the sides BC, CA, AB or their extensions at the points A', B', C'. Then*

$$\frac{\|GA'\|}{\|AA'\|} + \frac{\|GB'\|}{\|BB'\|} + \frac{\|GC'\|}{\|CC'\|} = 1. \tag{5.22}$$

The statement (5.22) has the form of a sum, therefore we call expressions (5.21), (5.22) *cyclic sums* [123]. Since we investigate ratios of the lengths of segments, the use of an affine system of coordinates is advantageous (Fig. 5.15).

First we will prove Euler's theorem by computer. Denoting $k_1 = \|GA'\|/\|AA'\|$, we can express a quotient of signed lengths $\|GA'\|/\|AA'\|$ as the ratio $k_1 = (GAA')$, etc. We get

$$\begin{aligned} k_1 &= \frac{\|GA'\|}{\|AA'\|} = (GAA') \Leftrightarrow k_1(A' - A) = A' - G. \\ k_2 &= \frac{\|GB'\|}{\|BB'\|} = (GBB') \Leftrightarrow k_2(B' - B) = B' - G, \\ k_3 &= \frac{\|GC'\|}{\|CC'\|} = (GCC') \Leftrightarrow k_3(C' - C) = C' - G. \end{aligned} \tag{5.23}$$

With the same choice of an affine coordinate system as in Fig. 5.15, denote

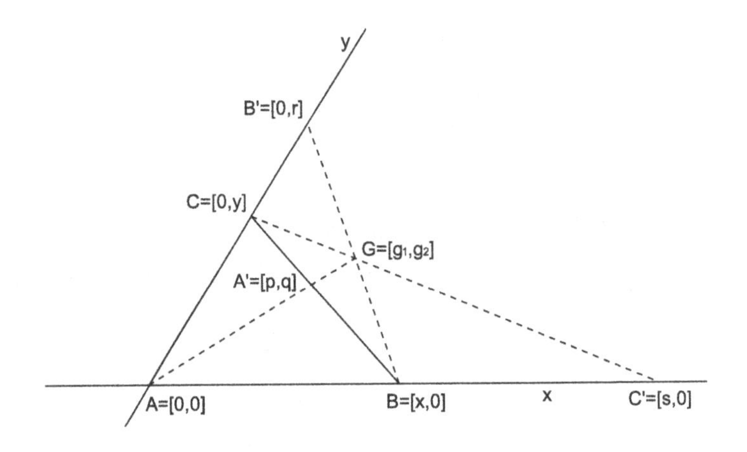

Fig. 5.15 Euler's theorem: $\frac{\|GA'\|}{\|AA'\|}+\frac{\|GB'\|}{\|BB'\|}+\frac{\|GC'\|}{\|CC'\|}=1$

$A=[0,0]$, $B=[x,0]$, $C=[0,y]$, $A'=[p,q]$, $B'=[0,r]$, $C'=[s,0]$, $G=[g_1,g_2]$.

From (5.23) we get the following relations:

$$k_1=(GAA')\Leftrightarrow h_1: k_1p-p+g_1=0,$$
$$h_2: k_1q-q+g_2=0,$$
$$k_2=(GBB')\Leftrightarrow h_3: -k_2x+g_1=0,$$
$$h_4: k_2r-r+g_2=0,$$
$$k_3=(GCC')\Leftrightarrow h_5: k_3s-s+g_1=0,$$
$$h_6: -k_3y+g_2=0.$$

Further

$A'\in BC\Leftrightarrow h_7: py+qx-xy=0.$

We want to prove that

$c: k_1+k_2+k_3=1.$

We will derive the formula (5.22). In the ideal $I=(h_1,h_2,\ldots,h_7)$ we eliminate variables p,q,r,s,g_1,g_2 and get

```
Use R::=Q[xypqrsg[1..2]k[1..3]];
I:=Ideal(k[1]p-p+g[1],k[1]q-q+g[2],-k[2]x+g[1],k[2]r-r+g[2],
```

```
k[3]s-s+g[1],-k[3]y+g[2],py+xq-xy);
Elim(p..g[2],I);
```

the polynomial $xy(k_1 + k_2 + k_3 - 1)$. Assuming that $xy \neq 0$, i.e., $A \neq B$, $A \neq C$, then the relation

$$k_1 + k_2 + k_3 = 1$$

follows. The verification

```
Use R::=Q[xypqrsg[1..2]k[1..3]t];
J:=Ideal(k[1]p-p+g[1],k[1]q-q+g[2],-k[2]x+g[1],k[2]r-r+g[2],
k[3]s-s+g[1],-k[3]y+g[2],py+xq-xy,xyt-1);
NF(k[1]+k[2]+k[3]-1,J);
```

with `NF=0` confirms that the relation (5.22) holds. The Euler's theorem is proved.

Remark 5.6. The formula (5.22) is equivalent to

$$\frac{\|AG\|}{\|AA'\|} + \frac{\|BG\|}{\|BB'\|} + \frac{\|CG\|}{\|CC'\|} = 2. \tag{5.24}$$

This follows from

$$\frac{\|AG\|}{\|AA'\|} + \frac{\|BG\|}{\|BB'\|} + \frac{\|CG\|}{\|CC'\|} =$$

$$\frac{\|AA'\| - \|GA'\|}{\|AA'\|} + \frac{\|BB'\| - \|GB'\|}{\|BB'\|} + \frac{\|CC'\| - \|GC'\|}{\|CC'\|} =$$

$$3 - (\frac{\|GA'\|}{\|AA'\|} + \frac{\|GB'\|}{\|BB'\|} + \frac{\|GC'\|}{\|CC'\|}) = 2.$$

In [10] it is written that the relation (5.24) was published in 1818 by J. D. Gergonne.

The following relation is due to L. Euler in 1780, see [36, 121]:

Given a triangle ABC and an arbitrary point G in the plane of a triangle which does not lie on the sides of ABC. Let the lines AG, BG, CG intersect the sides BC, CA, AB or their extensions at the points A', B', C'. Then

$$\frac{\|AG\|}{\|GA'\|} \cdot \frac{\|BG\|}{\|GB'\|} \cdot \frac{\|CG\|}{\|GC'\|} = \frac{\|AG\|}{\|GA'\|} + \frac{\|BG\|}{\|GB'\|} + \frac{\|CG\|}{\|GC'\|} + 2. \tag{5.25}$$

Here again $\|XY\|$ means the signed length of a segment XY.

There are many relations for a product of ratios of distances of the type on the left side (5.25). There exist formulas for a sum of ratios of lengths as well. (5.25) is one of few such formulas which connects a sum and product (see [50]). Let us derive the formula (5.25).

Denoting $\|AG\|/\|GA'\| = k_1$, $\|BG\|/\|GB'\| = k_2$, $\|CG\|/\|GC'\| = k_3$ and using the same notation as in the previous case $A = [0,0]$, $B = [x,0]$, $C = [0,y]$, $A' = [p,q]$, $B' = [0,r]$, $C' = [s,0]$, $G = [g_1,g_2]$ we get the following relations (Fig. 5.15):

$k_1(A' - G) = G - A \Leftrightarrow$
$h_1 : k_1(p - g_1) - g_1 = 0,\ h_2 : k_1(q - g_2) - g_2 = 0,$

$k_2(B' - G) = B - G \Leftrightarrow$
$h_3 : -k_2 g_1 - g_1 + x = 0,\ h_4 : k_2(g_2 - r) - g_2 = 0,$

$k_3(C' - G) = G - C \Leftrightarrow$
$h_5 : k_3(s - g_1) - g_1 = 0, h_6 : -k_3 g_2 - g_2 + y = 0.$

Further suppose that the points A', B, C are collinear:

A', B, C are collinear $\Leftrightarrow h_7 : py + qx - xy = 0.$

Elimination of variables p, q, r, s, g_1, g_2 in the ideal $I = (h_1, h_2, \ldots, h_7)$ gives

```
Use R::=Q[xypqrsg[1..2]k[1..3]];
I:=Ideal(k[1](p-g[1])-g[1],k[1](q-g[2])-g[2],-k[2]g[1]-g[1]+x,
k[2](g[2]-r)-g[2],k[3](s-g[1])-g[1],-k[3]g[2]-g[2]+y,py+qx-xy);
Elim(p..g[2],I);
```

the only polynomial $xy(k_1k_2k_3 - k_1 - k_2 - k_3 - 2)$. Assuming that $xy \neq 0$ we get the condition

$$k_1k_2k_3 - k_1 - k_2 - k_3 - 2 = 0$$

which is the relation (5.25). The verification

```
Use R::=Q[xypqrsg[1..2]k[1..3]t];
J:=Ideal(k[1](p-g[1])-g[1],k[1](q-g[2])-g[2],-k[2]g[1]-g[1]+x,
k[2](g[2]-r)-g[2],k[3](s-g[1])-g[1],-k[3]g[2]-g[2]+y,xyt-1,
py+xq-xy);
NF(k[1]k[2]k[3]-k[1]-k[2]-k[3]-2,J);
0
```

confirms that (5.25) is true.

Now we will prove (5.25) classically by the area principle [22, 122]. Denote the signed areas of the triangles ABG, BCG, CAG by $p = \|ABG\|$, $q = \|BCG\|$, $r = \|CAG\|$ respectively (Fig. 5.16). It holds

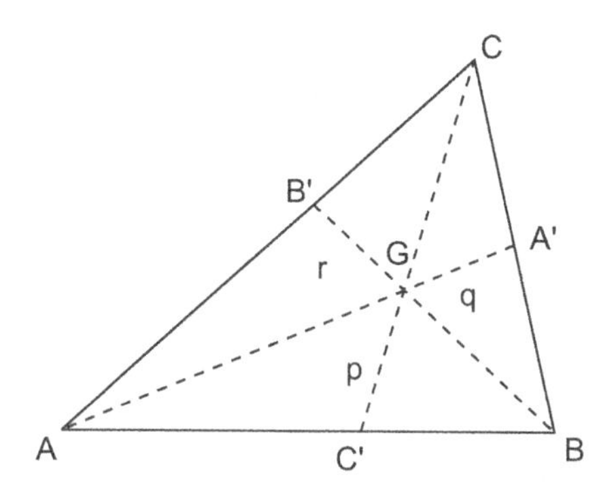

Fig. 5.16 Euler's theorem — classical proof

$$\frac{\|AG\|}{\|GA'\|} = \frac{\|AA'\| - \|GA'\|}{\|GA'\|} = \frac{\|AA'\|}{\|GA'\|} - 1 = \frac{p+q+r}{q} - 1 = \frac{p+r}{q}. \quad (5.26)$$

Analogously

$$\frac{\|BG\|}{\|GB'\|} = \frac{p+q}{r}, \ \frac{\|CG\|}{\|GC'\|} = \frac{q+r}{p}. \quad (5.27)$$

Hence

$$\frac{\|AG\|}{\|GA'\|} \cdot \frac{\|BG\|}{\|GB'\|} \cdot \frac{\|CG\|}{\|GC'\|} = \frac{p+r}{q} \cdot \frac{p+q}{r} \cdot \frac{q+r}{p}$$

$$= \frac{q^2r + r^2q + r^2p + p^2r + p^2q + q^2p + 2pqr}{pqr}$$

$$= \frac{p+r}{q} + \frac{p+q}{r} + \frac{q+r}{p} + 2 = \frac{\|AG\|}{\|GA'\|} + \frac{\|BG\|}{\|GB'\|} + \frac{\|CG\|}{\|GC'\|} + 2.$$

The Euler's theorem is thus proved.

By means of (5.26) and (5.27) we will easily prove the formula (5.22) of the Euler's theorem in a classical way. Namely it holds

$$\frac{\|GA'\|}{\|AA'\|} = \frac{q}{p+q+r}, \ \frac{\|GB'\|}{\|BB'\|} = \frac{r}{p+q+r}, \ \frac{\|GC'\|}{\|CC'\|} = \frac{p}{p+q+r}$$

and from here

$$\frac{\|GA'\|}{\|AA'\|} + \frac{\|GB'\|}{\|BB'\|} + \frac{\|GC'\|}{\|CC'\|} = \frac{q}{p+q+r} + \frac{r}{p+q+r} + \frac{p}{p+q+r} = 1.$$

5.3.1 *Spatial analogue of Euler's theorem*

We will study an analogue of the Euler's relation (5.22) in the space. We prove the following theorem:

Theorem 5.7. *Given a tetrahedron $ABCD$ and an arbitrary point O. Denote by A', B', C', D' intersections of the lines AO, BO, CO, DO with the planes BCD, ACD, ABD, ABC respectively. Then*

$$\frac{\|OA'\|}{\|AA'\|} + \frac{\|OB'\|}{\|BB'\|} + \frac{\|OC'\|}{\|CC'\|} + \frac{\|OD'\|}{\|DD'\|} = 1. \tag{5.28}$$

Proof by computer is as follows.

Since the relation (5.28) is an affine invariant, we use the affine coordinate system so that $A = [0,0,0]$, $B = [x,0,0]$, $C = [0,y,0]$, $D = [0,0,z]$ (Fig. 5.17). Denote $O = [o_1,o_2,o_3]$, $A' = [a_1,a_2,a_3]$, $B' = [0,b_2,b_3]$,

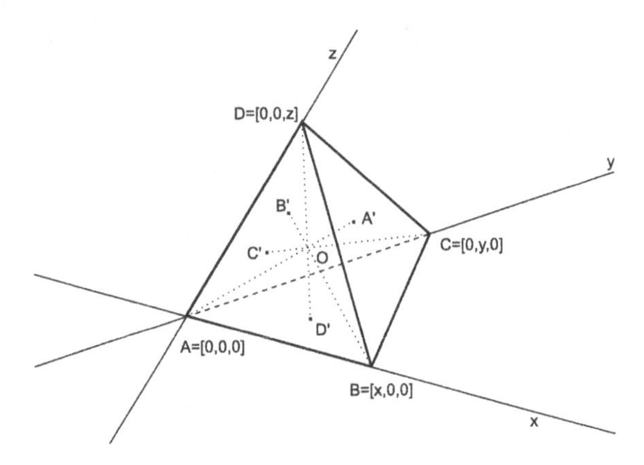

Fig. 5.17 Spatial analogue of Euler's theorem

$C' = [c_1,0,c_3]$, $D' = [d_1,d_2,0]$ and let

$$\frac{\|OA'\|}{\|AA'\|} = k_1, \quad \frac{\|OB'\|}{\|BB'\|} = k_2, \quad \frac{\|OC'\|}{\|CC'\|} = k_3, \quad \frac{\|OD'\|}{\|DD'\|} = k_4. \tag{5.29}$$

From (5.29) we get the following hypotheses:

$k_1(A'-A) = A'-O \Leftrightarrow$
$h_1: k_1a_1 - a_1 + o_1 = 0,\ h_2: k_1a_2 - a_2 + o_2 = 0,\ h_3: k_1a_3 - a_3 + o_3 = 0,$

$k_2(B'-B) = (B'-O) \Leftrightarrow$
$h_4: k_2(-x) + o_1 = 0,\ h_5: k_2b_2 - b_2 + o_2 = 0,\ h_6: k_2b_3 - b_3 + o_3 = 0,$

$k_3(C'-C) = C'-O \Leftrightarrow$
$h_7: k_3c_1 - c_1 + o_1 = 0,\ h_8: k_3(-y) + o_2 = 0,\ h_9: k_3c_3 - c_3 + o_3 = 0,$

$k_4(D'-D) = D'-O \Leftrightarrow$
$h_{10}: k_4d_1 - d_1 + o_1 = 0,\ h_{11}: k_4d_2 - d_2 + o_2 = 0,\ h_{12}: k_4(-z) + o_3 = 0$

and

$A' \in BCD \Leftrightarrow$
$h_{13}: xza_2 + xya_3 + yza_1 - xyz = 0.$

We will derive the statement (5.28). In the ideal $I = (h_1, h_2, \ldots, h_{13})$ we eliminate variables $a_1, a_2, a_3, b_2, b_3, c_1, c_3, d_1, d_2, o_1, o_2, o_3$

```
Use R::=Q[xyza[1..3]b[1..3]c[1..3]d[1..3]o[1..3]k[1..4]];
I:=Ideal(k[1]a[1]-a[1]+o[1],k[1]a[2]-a[2]+o[2],k[1]a[3]-a[3]
+o[3],k[2](-x)+o[1],k[2]b[2]-b[2]+o[2],k[2]b[3]-b[3]+o[3],
k[3]c[1]-c[1]+o[1],k[3](-y)+o[2],k[3]c[3]-c[3]+o[3],k[4]d[1]
-d[1]+o[1],k[4]d[2]-d[2]+o[2],k[4](-z)+o[3],xza[2]+xya[3]+
yza[1]-xyz);
Elim(a[1]..o[3],I);
```

and get $xyz(k_1 + k_2 + k_3 + k_4 - 1) = 0$. If we assume that $xyz \neq 0$, i.e., the points A, B, C, D are distinct, then the relation (5.28) follows. Theorem 5.7 is proved.

Now we will show the analogue of the Euler's relation (5.25) in the space (Fig. 5.17).

Assume that a tetrahedron $ABCD$ and an arbitrary point O are given. Let A', B', C', D' be the intersections of the lines AO, BO, CO, DO with the planes BCD, ACD, ABD, ABC respectively. Denote

$$k_1 = \frac{\|AO\|}{\|OA'\|},\ k_2 = \frac{\|BO\|}{\|OB'\|},\ k_3 = \frac{\|CO\|}{\|OC'\|},\ k_4 = \frac{\|DO\|}{\|OD'\|}.$$

Find a relation that holds for the ratios k_1, k_2, k_3, k_4.

On the base of equalities $k_1(A'-O) = O-A$, $k_2(B'-O) = O-B$, $k_3(C'-O) = O-C$, $k_4(D'-O) = O-D$ we get

$k_1(A'-O) = O-A \Leftrightarrow$

$h_1 : k_1(a_1 - o_1) - o_1 = 0,\ h_2 : k_1(a_2 - o_2) - o_2 = 0,\ h_3 : k_1(a_3 - o_3) - o_3 = 0,$

$k_2(B' - O) = O - B \Leftrightarrow$
$h_4 : k_2(b_1 - o_1) - o_1 + x = 0,\ h_5 : k_2(b_2 - o_2) - o_2 = 0,\ h_6 : k_2(b_3 - o_3) - o_3 = 0,$

$k_3(C' - O) = O - C \Leftrightarrow$
$h_7 : k_3(c_1 - o_1) - o_1 = 0,\ h_8 : k_3(c_2 - o_2) - o_2 + y = 0,\ h_9 : k_3(c_3 - o_3) - o_3 = 0,$

$k_4(D' - O) = O - D \Leftrightarrow$
$h_{10} : k_4(d_1 - o_1) - o_1 = 0,\ h_{11} : k_4(d_2 - o_2) - o_2 = 0,\ h_{12} : k_4(d_3 - o_3) - o_3 + z = 0.$

Conditions $A' \in BCD$, $B' \in ACD$, $C' \in ABD$, $D' \in ABC$ imply

$A' \in BCD \Leftrightarrow h_{13} : xza_2 + xya_3 + yza_1 - xyz = 0,$

$B' \in ACD \Leftrightarrow h_{14} : b_1 = 0,$

$C' \in ABD \Leftrightarrow h_{15} : c_2 = 0,$

$D' \in ABC \Leftrightarrow h_{16} : d_3 = 0.$

The elimination of all variables except $x, y, z, k_1, k_2, k_3, k_4$ in the ideal $I = (h_1, h_2, \ldots, h_{16})$

```
Use R::=Q[xyza[1..3]b[1..3]c[1..3]d[1..3]o[1..3]k[1..4]];
I:=Ideal(k[1](a[1]-o[1])-o[1],k[1](a[2]-o[2])-o[2],k[1](a[3]-
o[3])-o[3],k[2](b[1]-o[1])-o[1]+x,k[2](b[2]-o[2])-o[2],k[2]
(b[3]-o[3])-o[3],k[3](c[1]-o[1])-o[1],k[3](c[2]-o[2])-o[2]+y,
k[3](c[3]-o[3])-o[3],k[4](d[1]-o[1])-o[1],k[4](d[2]-o[2])-o[2],
k[4](d[3]-o[3])-o[3]+z,xza[2]+xya[3]+yza[1]-xyz,b[1],c[2],d[3]);
Elim(a[1]..o[3],I);
```

gives the relation

$$k_1k_2k_3k_4 = (k_1k_2 + k_1k_3 + k_2k_3 + k_1k_4 + k_2k_4 + k_3k_4) + 2(k_1 + k_2 + k_3 + k_4) + 3. \tag{5.30}$$

Under the assumption $xyz \neq 0$, i.e., the vertices A, B, C, D of a tetrahedron are distinct, the polynomial in (5.30) belongs to the ideal I. We proved the theorem:

Theorem 5.8. *Given a tetrahedron $ABCD$ and an arbitrary point O. Let A', B', C', D' be the intersections of the lines AO, BO, CO, DO with the planes BCD, ACD, ABD, ABC respectively. Denote*

$$k_1 = \frac{\|AO\|}{\|OA'\|},\ k_2 = \frac{\|BO\|}{\|OB'\|},\ k_3 = \frac{\|CO\|}{\|OC'\|},\ k_4 = \frac{\|DO\|}{\|OD'\|}.$$

Then for k_1, k_2, k_3, k_4 the equality (5.30) holds.

Remark 5.7. It seems that Theorem 5.8 above has not been published yet.

5.4 Routh's theorem

The following theorem, which is a generalization of the theorem of Ceva was published in 1896 by the English mathematician E. J. Routh, see [113, 62]:

Theorem 5.9 (Routh). *Given a triangle ABC and three points A', B', C' on its sides BC, AC, AB or their extensions respectively. Denote by $l_1 = (BCA')$, $l_2 = (CAB')$ and $l_3 = (ABC')$ the ratios of points A', B', C' respectively. Then the lines AA', BB', CC' form a triangle GHI (Fig. 5.18), for which the ratio of the areas of triangles GHI and ABC equals*

$$\frac{\|GHI\|}{\|ABC\|} = \frac{(l_1l_2l_3+1)^2}{(l_1l_2-l_1+1)(l_2l_3-l_2+1)(l_3l_1-l_3+1)}. \tag{5.31}$$

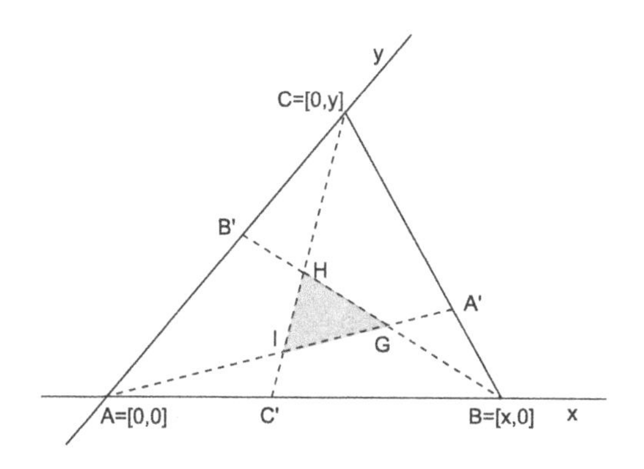

Fig. 5.18 Routh's theorem

A few proofs of this theorem are known. We will try, similarly as in the case of Ceva's theorem, to *derive* the formula (5.31). Since we deal with the problem of affine geometry (ratios of points, ratios of areas of triangles), we will work in an affine coordinate system.

Denote $A = [0,0]$, $B = [x,0]$, $C = [0,y]$, $A' = [p,q]$, $B' = [0,r]$, $C' = [s,0]$, $G = [g_1,g_2]$, $H = [h_1,h_2]$, $I = [i_1,i_2]$ (Fig. 5.18).
Instead of $l_1 = (BCA')$ we can write $l_1(A'-C) = A'-B$ and analogously

$l_2(B' - A) = B' - C$, $l_3(C' - B) = C' - A$. This implies

$$l_1 = (BCA') \Leftrightarrow h_1 : l_1(q - y) - q = 0,$$

$$h_2 : l_1 p - p + x = 0,$$

$$l_2 = (CAB') \Leftrightarrow h_3 : l_2 r - r + y = 0,$$

$$l_3 = (ABC') \Leftrightarrow h_4 : l_3(s - x) - s = 0.$$

Further

$$G \in AA' \Leftrightarrow h_5 : qg_1 - pg_2 = 0,$$

$$G \in BB' \Leftrightarrow h_6 : rg_1 + xg_2 - xr = 0,$$

$$H \in BB' \Leftrightarrow h_7 : rh_1 + xh_2 - xr = 0,$$

$$H \in CC' \Leftrightarrow h_8 : yh_1 + sh_2 - sy = 0,$$

$$I \in AA' \Leftrightarrow h_9 : qi_1 - pi_2 = 0,$$

$$I \in CC' \Leftrightarrow h_{10} : yi_1 + si_2 - sy = 0.$$

In the affine system let us put $x = y = 1$. Then for the ratio f of the areas of the triangles GHI and ABC

$$h_{11} : g_1 h_2 + h_1 i_2 + g_2 i_1 - h_2 i_1 - g_1 i_2 - g_2 h_1 - f = 0$$

holds.

Elimination of variables $x, y, p, q, r, s, g_1, g_2, h_1, h_2, i_1, i_2$ in the ideal $I = (h_1, h_2, \ldots, h_{11}) \cup \{x - 1\} \cup \{y - 1\}$ gives

```
Use R::=Q[xypqrsg[1..2]h[1..2]i[1..2]l[1..3]f];
I:=Ideal(l[1](q-y)-q,l[1]p-p+x,l[2]r-r+y,l[3](s-x)-s,qg[1]-
pg[2],rg[1]+xg[2]-xr,rh[1]+xh[2]-xr,yh[1]+sh[2]-sy,qi[1]-pi[2],
yi[1]+si[2]-sy,x-1,y-1,f-g[1]h[2]-h[1]i[2]-g[2]i[1]+h[2]i[1]+
g[1]i[2]+g[2]h[1]);
Elim(x..i[2],I);
```

the equation

$$-f(l_2 l_3 - l_2 + 1)(l_1 l_3 - l_3 + 1)(l_1 l_2 - l_1 + 1) + (l_1 l_2 l_3 + 1)^2 = 0,$$

which is the relation (5.31).

Remark 5.8. If we denote

$$l_1 = \frac{\|BA'\|}{\|A'C\|}, \; l_2 = \frac{\|CB'\|}{\|B'A\|}, \; l_3 = \frac{\|AC'\|}{\|C'B\|}$$

then (5.31) is equivalent to

$$\frac{\|GHI\|}{\|ABC\|} = \frac{(l_1 l_2 l_3 - 1)^2}{(l_1 l_2 + l_1 + 1)(l_2 l_3 + l_2 + 1)(l_3 l_1 + l_3 + 1)}. \tag{5.32}$$

Namely it suffices to write $-l_1, -l_2, -l_3$ instead of l_1, l_2, l_3 in (5.31) since

$$(BCA') = -\frac{\|BA'\|}{\|A'C\|},$$

etc.

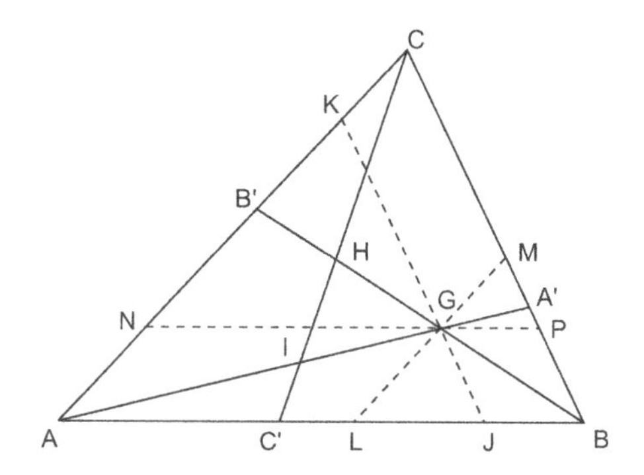

Fig. 5.19 Routh's theorem — classical proof

Now we shall show a classical proof of Routh's theorem. We retain the method used by J. S. Kline and D. J. Velleman [62]. We prove the Routh's theorem in the form (5.32).

By constructing parallels to the sides of a triangle ABC through the point G (Fig. 5.19), we have

$$\frac{\|JG\|}{\|GK\|} = \frac{\|BA'\|}{\|A'C\|} = l_1,$$

and from here $\|GK\| = \|JG\|/l_1$. From the similarity of triangles LJG and GPM we get

$$\frac{\|PM\|}{\|JG\|} = \frac{\|MG\|}{\|GL\|} = \frac{\|CB'\|}{\|B'A\|} = l_2$$

and $\|PM\| = l_2 \cdot \|JG\|$.

Further

$\|BC\| = \|BP\| + \|PM\| + \|MC\| = \|JG\| + \|PM\| + \|GK\| =$

$\|JG\| + l_2 \cdot \|JG\| + 1/l_1 \cdot \|JG\|,$

hence

$$\frac{\|JG\|}{\|BC\|} = \frac{l_1}{l_1 l_2 + l_1 + 1}. \tag{5.33}$$

Triangles LJG and ABC are similar with the ratio (5.33). Ratios of the lengths of altitudes to the sides LJ and AB are alike. This implies that the ratio of the area of ABG to the area of ABC equals (5.33). Thus we can write

$$\frac{\|ABG\|}{\|ABC\|} = \frac{l_1}{l_1 l_2 + l_1 + 1}.$$

Analogously for areas of the triangles BCH and CAI we get

$$\frac{\|BCH\|}{\|ABC\|} = \frac{l_2}{l_2 l_3 + l_2 + 1} \quad \text{and} \quad \frac{\|CAI\|}{\|ABC\|} = \frac{l_3}{l_3 l_1 + l_3 + 1}.$$

Whence

$$\frac{\|GHI\|}{\|ABC\|} = 1 - \frac{l_1}{l_1 l_2 + l_1 + 1} - \frac{l_2}{l_2 l_3 + l_2 + 1} - \frac{l_3}{l_3 l_1 + l_3 + 1}$$

$$= \frac{(l_1 l_2 l_3 - 1)^2}{(l_1 l_2 + l_1 + 1)(l_2 l_3 + l_2 + 1)(l_3 l_1 + l_3 + 1)}$$

which is the relation (5.32). The theorem is proved.

Remark 5.9.
1) A special case of Routh's theorem is the theorem of Ceva. Namely, if we put in (5.31) $l_1 l_2 l_3 = -1$ then the area of the triangle GHI vanishes and the lines AA', BB', CC' are concurrent.

2) If the points A', B', C' divide the sides of a triangle ABC in the same ratio then $l_1 = l_2 = l_3 = l$ and the formula (5.31) has the form

$$\frac{\|GHI\|}{\|ABC\|} = \frac{(l + 1)^3}{l^3 + 1}. \tag{5.34}$$

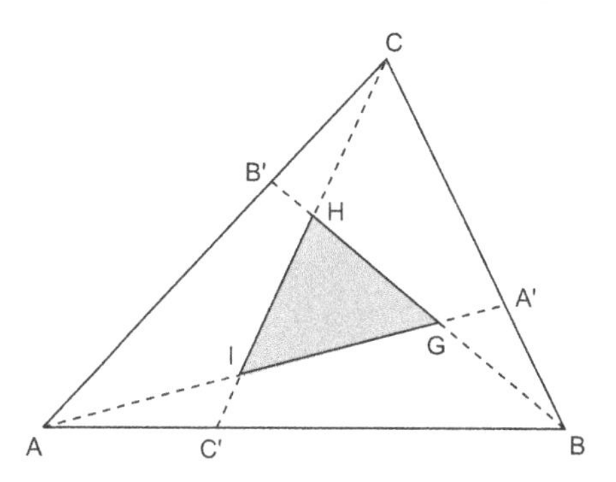

Fig. 5.20 Routh's theorem for values $l_1 = l_2 = l_3 = -1/2$

Dividing sides of a triangle into three like segments then $l_1 = l_2 = l_3 = -1/2$ and from (5.34) we get that the area of the triangle GHI equals 1/7 of the area of ABC (Fig. 5.20).

We conclude this section with an elegant proof of this statement, which is given by R. B. Nelsen in Proofs Without Words II [80]. We will give his proof also without words (Fig. 5.21).

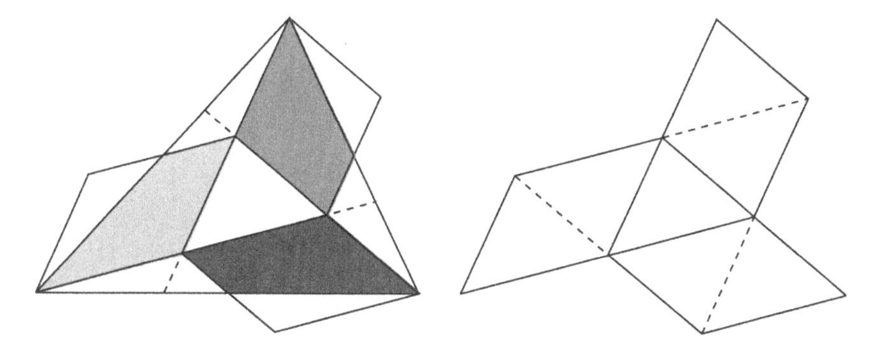

Fig. 5.21 Proof without words: Routh's theorem for values $l_1 = l_2 = l_3 = -1/2$

Chapter 6

Petr–Douglas–Neumann theorem

The Petr–Douglas–Neumann theorem (PDN theorem) has a rich history [72, 73, 88, 91, 46]. Its name is closely connected with the Czech mathematician Karel Petr who first published this theorem in 1905 (see [97]). Perhaps, because of the fact that the work [97] was written in Czech (although two years later a German version [98] appeared), it was assumed for a long time that the authors of this theorem were J. Douglas [30, 31] and B. H. Neumann [81], who published their works in the forties of the last century. The PDN theorem, also known as the Petr's theorem, is mentioned in [77, 74]. The name "Petr–Douglas–Neumann theorem" was given by H. Martini in 1996 in his paper [73].

There are many properties of polygons in elementary geometry in the plane and space which are connected with the PDN theorem even though we do not realize this. The list of problems connected with the PDN theorem would deserve a special publication.

In this chapter we will give basic features of the PDN theorem mainly from the geometry of a triangle, quadrilateral and pentagon. A large space is devoted to the best known special case of the PDN theorem which is the theorem of Napoleon.

Our approach is in line with that adopted in the other chapters. We will solve (prove, derive or discover) the problem by computer followed by the classical method. At the end of this chapter we will study a spatial version of the PDN theorem.

6.1 Napoleon's theorem

We will deal with a theorem which is called *Napoleon's theorem.* This theorem is often ascribed to the well-known emperor Napoleon Bonaparte

[41, 119], although there are doubts about his knowing enough geometry to prove it. The Napoleon's theorem reads:

Theorem 6.1 (Napoleon). *On the sides of an arbitrary triangle construct equilateral triangles (all outwardly or all inwardly). Then their centers form an equilateral triangle (Fig. 6.1).*

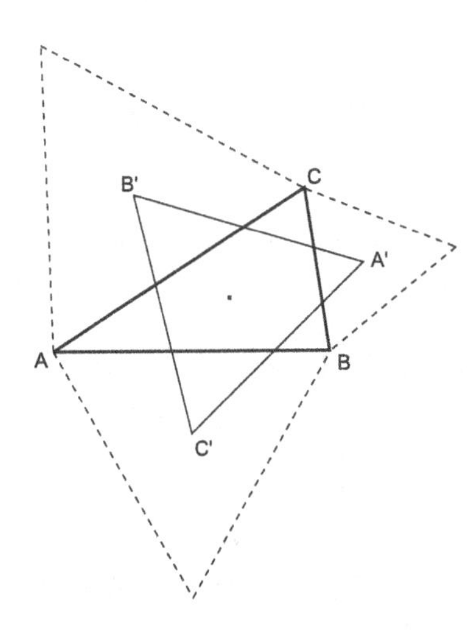

Fig. 6.1 Napoleon's theorem — the triangle $A'B'C'$ is equilateral

First we will try to "discover" this theorem by computer. Over the sides of a triangle ABC we construct *similar isosceles* triangles ABC', BCA', CAB'. We search for such isosceles triangles whose vertices A', B', C' form an equilateral triangle.

Choose a Cartesian coordinate system so that $A = [0,0]$, $B = [a,0]$, $C = [b,c]$, $A' = [k_1,k_2]$, $B' = [l_1,l_2]$, $C' = [m_1,m_2]$ and on the sides of ABC outwardly construct *arbitrary* similar isosceles triangles ABC', BCA', CAB' (Fig. 6.2).

The problem of this task is to express the fact "to lie outwardly" by algebraic *equations*. Methods of proofs we use are based on the Hilbert's theorem Nullstellensatz which holds in an algebraic closed field, for instance, in the field of complex numbers. However, it is well-known that complex numbers cannot be ordered. That is why we cannot use algebraic inequalities to express a half-plane, etc. In this case we shall apply the method which is due to D. Wang [134].

The vertex A' is the endpoint of a vector whose initial point is in the center of BC with the length $v|BC|$, where v is an arbitrary real number, and the same direction as the vector $B - C$ rotated by the angle 90° in a positive sense. We will denote $A' - (B+C)/2 = v \cdot \mathrm{rot}(B - C)$. For the coordinates k_1, k_2 of the point A' the relation

$$(k_1 - \frac{a+b}{2}, k_2 - \frac{c}{2}) = v(c, a-b) \tag{6.1}$$

holds. Similarly, for the vertices B', C' we get

$$(l_1 - \frac{b}{2}, l_2 - \frac{c}{2}) = v(-c, b), \quad (m_1 - \frac{a}{2}, m_2) = v(0, -a). \tag{6.2}$$

The number v is the same on all three sides because of the similarity of the triangles ABC', BCA', CAB'. We are looking for such a real number v for which the triangle $A'B'C'$ becomes equilateral. For the coordinates of the vertices A', B', C' we get from (6.1), (6.2) the following conditions:

$A' - (B+C)/2 = v \cdot \mathrm{rot}(B - C) \Leftrightarrow$
$h_1 : 2k_1 - a - b - 2vc = 0,\ h_2 : 2k_2 - c - 2va + 2vb = 0,$

$B' - (A+C)/2 = v \cdot \mathrm{rot}(C - A) \Leftrightarrow$
$h_3 : 2l_1 - b + 2vc = 0,\ h_4 : 2l_2 - c - 2vb = 0,$

$C' - (A+B)/2 = v \cdot \mathrm{rot}(A - B) \Leftrightarrow$
$h_5 : 2m_1 - a = 0,\ h_6 : m_2 + va = 0.$

As a conclusion we require the conditions $|A'B'| = |A'C'|$ and $|A'B'| = |B'C'|$ to be fulfilled, that is,

$$c_1 : (k_1 - l_1)^2 + (k_2 - l_2)^2 - (m_1 - k_1)^2 - (m_2 - k_2)^2 = 0,$$

$$c_2 : (k_1 - l_1)^2 + (k_2 - l_2)^2 - (m_1 - l_1)^2 - (m_2 - l_2)^2 = 0.$$

First we need the validity of the equality $|A'B'| = |A'C'|$. Let $I = (h_1, h_2, \ldots, h_6)$. In the ideal $J = I \cup \{c_1\}$ we eliminate dependent variables $k_1, k_2, l_1, l_2, m_1, m_2$.

```
Use R::=Q[k[1..2]l[1..2]m[1..2]abcv];
J:=Ideal(2k[1]-a-b-2vc,2k[2]-c-2va+2vb,2l[1]-b+2vc,2l[2]-c-2vb,
2m[1]-a,m[2]+va,(k[1]-l[1])^2+(k[2]-l[2])^2-(m[1]-k[1])^2-(m[2]
-k[2])^2);
Elim(k[1]..m[2],J);
```

The resulting elimination ideal gives the only condition

$$(12v^2 - 1)(a^2 - b^2 - c^2) = 0,$$

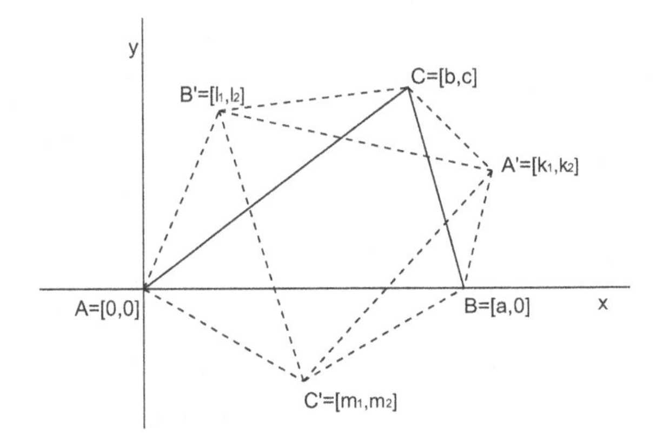

Fig. 6.2 Napoleon's theorem — proof by computer

from which we get:

a) $a^2 = b^2 + c^2$, i.e. $|AB| = |AC|$. Then the triangle ABC is isosceles and v is an *arbitrary* number

or

b) $v = \sqrt{3}/6$ or $v = -\sqrt{3}/6$ and the triangle ABC is arbitrary.

If a triangle ABC is isosceles then the theorem obviously holds for every v. However, we are more interested in the values $v = \sqrt{3}/6$ and $v = -\sqrt{3}/6$. Let us denote

$c_3 : 12v^2 - 1 = 0$

and verify the statement for these values and an arbitrary triangle ABC. We will show that the polynomial c_1 belongs to the ideal $K = I \cup \{c_3\}$. In CoCoA we enter

```
Use R::=Q[k[1..2]l[1..2]m[1..2]abcv];
K:=Ideal(2k[1]-a-b-2vc,2k[2]-c-2va+2vb,2l[1]-b+2vc,2l[2]-c-2vb,
2m[1]-a,m[2]+va,12v^2-1);
NF((k[1]-l[1])^2+(k[2]-l[2])^2-(m[1]-l[1])^2-(m[2]-l[2])^2,K);
```

and get the result 0. This means that for the values $v = \sqrt{3}/6$ or $v = -\sqrt{3}/6$ the equality $|A'B'| = |A'C'|$ holds.

Analogously we prove that for the same values $v = \sqrt{3}/6$ or $v = -\sqrt{3}/6$ the equality $|A'B'| = |B'C'|$ holds. Hence $|A'B'| = |B'C'| = |A'C'|$.

For $v = \sqrt{3}/6$ we acquire the *outer Napoleon triangle* $A'B'C'$, for $v = -\sqrt{3}/6$ we get the *inner Napoleon triangle* $A''B''C''$ (Fig. 6.3). The values

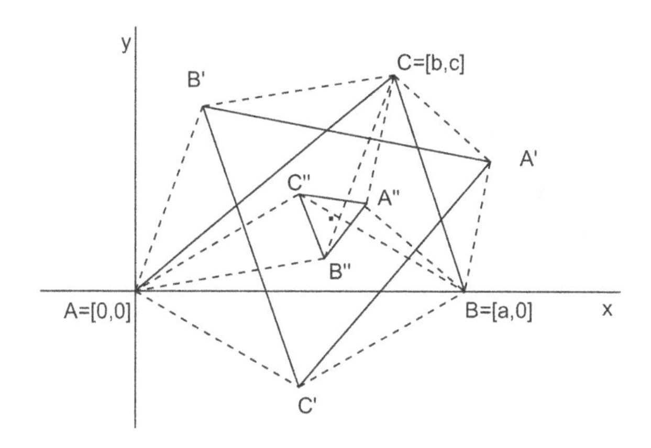

Fig. 6.3 Outer and inner Napoleon triangles

$v = \sqrt{3}/6$ and $v = -\sqrt{3}/6$ correspond to isosceles triangles with the angle $2\pi/3$ which were constructed outwardly and inwardly over the sides of a triangle ABC respectively. We proved (and rediscovered) the theorem of Napoleon.

Remark 6.1. The situation in Fig. 6.1 is often called the *Napoleon configuration.* This configuration has a lot of interesting properties. Some of them are discussed here.

Napoleon's triangles have the following properties [26]:

1) The sum of the signed areas of Napoleon's triangles $A'B'C'$ and $A''B''C''$ equals the area of a triangle ABC.

2) A triangle ABC and Napoleon triangles $A'B'C'$ and $A''B''C''$ have a common centroid.

First we prove the preceding statement. Using the same notation as by the proof of Napoleon's theorem 6.1, suppose that for the outer Napoleon triangle $A'B'C'$ the hypotheses $h_1, h_2, \ldots, h_6, c_3$ hold. Let us denote $A'' = [n_1, n_2]$, $B'' = [o_1, o_2]$, $C'' = [p_1, p_2]$ (Fig. 6.3). For the points A'', B'', C'', similarly as for the points A', B', C', the following relations hold:

$A'' - (B+C)/2 = v \cdot \mathrm{rot}(C-B) \Leftrightarrow$
$h_7 : 2n_1 - a - b + 2vc = 0,\ h_8 : 2n_2 - c + 2va - 2vb = 0,$

$B' - (A+C)/2 = v \cdot \mathrm{rot}(C-A) \Leftrightarrow$
$h_9 : 2o_1 - b - 2vc = 0,\ h_{10} : 2o_2 - c + 2vb = 0,$

$C' - (A+B)/2 = v \cdot \mathrm{rot}(A-B) \Leftrightarrow$
$h_{11} : 2p_1 - a = 0,\ h_{12} : p_2 - va = 0.$

Denote the signed areas of the triangles ABC, $A'B'C'$, $A''B''C''$ by f, g, h respectively. Then for the area f

$h_{13} : 2f - ac = 0,$

and for the areas g, h

$$h_{14} : g - \frac{1}{2}\begin{vmatrix} k_1 & k_2 & 1 \\ l_1 & l_2 & 1 \\ m_1 & m_2 & 1 \end{vmatrix} = 0, \quad h_{15} : h - \frac{1}{2}\begin{vmatrix} n_1 & n_2 & 1 \\ o_1 & o_2 & 1 \\ p_1 & p_2 & 1 \end{vmatrix} = 0.$$

We search for relations between the areas f, g, h. In the ideal $L = (h_1, \ldots, h_{15}, c_3)$ we eliminate all variables except f, g, h. CoCoA returns

```
Use R::=Q[k[1..2]l[1..2]m[1..2]n[1..2]o[1..2]p[1..2]abcvfgh];
L:=Ideal(2k[1]-a-b-2vc,2k[2]-c-2va+2vb,2l[1]-b+2vc,2l[2]-c-2vb,
2m[1]-a,m[2]+va,2n[1]-a-b+2vc,2n[2]-c+2va-2vb,2o[1]-b-2vc,2o[2]
-c+2vb,2p[1]-a,p[2]-va,2f-ac,2g-(k[1]l[2]+l[1]m[2]+m[1]
k[2]-m[1]l[2]-k[1]m[2]-k[2]l[1]),2h-(n[1]o[2]+o[1]p[2]+p[1]n[2]
-p[1]o[2]-n[1]p[2]-n[2]o[1]),12v^2-1);
Elim(k[1]..v,L);
```

the relation $f = g + h$. A verification gives `NF(f-g-h,L)=0` and the first property is proven.

To prove the second property, we will show even more, namely that *all* triangles which are analogous to Napoleon triangles for an *arbitrary* v (i.e. not only for the values $v = \pm\sqrt{3}/6$) have a common centroid. It holds (Fig. 6.4):

Let $A'B'C'$ be vertices of similar isosceles triangles which are erected on the sides of a triangle ABC (all outwardly or all inwardly). Then the triangles A, B, C and $A'B'C'$ have a common centroid.

It is well-known that for the centroid T of ABC the equality $T = 1/3(A + B + C)$ holds. We prove that $A + B + C = A' + B' + C'$ which is expressed by the equations

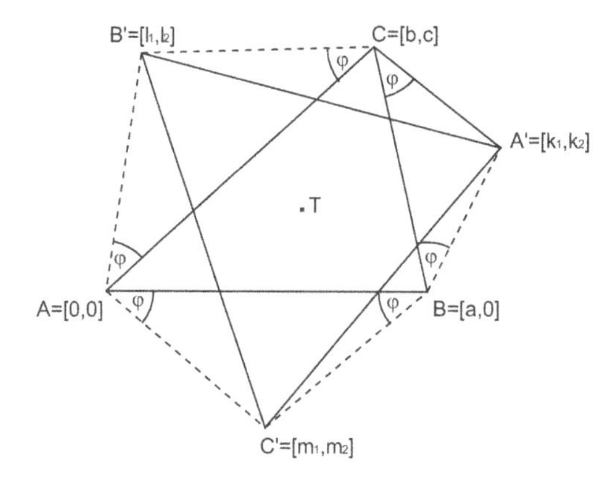

Fig. 6.4 Triangles A, B, C and $A'B'C'$ have a common centroid T

$z_1 : a + b - (k_1 + l_1 + m_1) = 0$ and $z_2 : c - (k_2 + l_2 + m_2) = 0.$

We show that the conclusion polynomials z_1, z_2 belong to the ideal $J = (h_1, h_2, \ldots, h_6)$. For z_1 we enter

```
Use R::=Q[k[1..2]l[1..2]m[1..2]abcvt];
J:=Ideal(2k[1]-a-b-2vc,2k[2]-c-2va+2vb,2l[1]-b+2vc,2l[2]-c-2vb,
2m[1]-a,m[2]+va);
NF(a+b-(k[1]+l[1]+m[1]),J);
```

and get the response 0. Analogously we prove the statement for z_2. Thus the second property is proven.

Now we will prove Napoleon's theorem in a classical way. We carry out the proof for the outer Napoleon triangle by [40] (see also [125]).

Denote $a = |BC|$, $b = |CA|$, $c = |AB|$ and apply the law of cosines to the triangle $AB'C'$ in Fig. 6.5. Then

$$|B'C'|^2 = |AB'|^2 + |AC'|^2 - 2|AB'||AC'|\cos(\alpha + 60^\circ). \qquad (6.3)$$

Since

$|AB'| = b/\sqrt{3}$, $|AC'| = c/\sqrt{3}$ and $\cos(\alpha + 60^\circ) = \frac{1}{2}\cos\alpha - \frac{\sqrt{3}}{2}\sin\alpha$,

then substitution into (6.3) gives

$$|B'C'|^2 = \frac{b^2}{3} + \frac{c^2}{3} - \frac{2bc}{3}\left(\frac{1}{2}\cos\alpha - \frac{\sqrt{3}}{2}\sin\alpha\right).$$

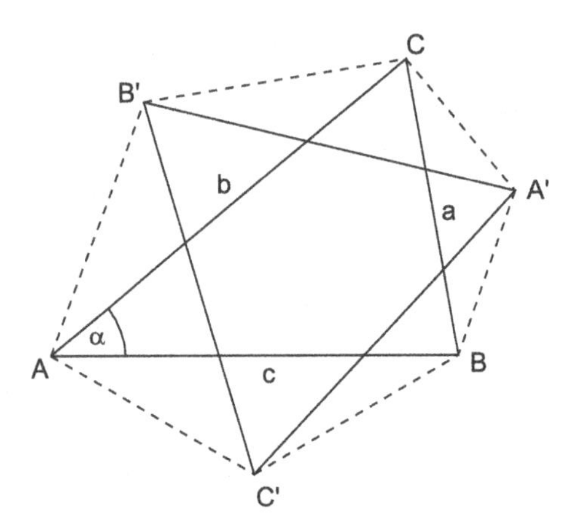

Fig. 6.5 Napoleon's theorem — classical proof

Furthermore, since $a^2 = b^2 + c^2 - 2bc\cos\alpha$ then

$$|B'C'|^2 = \frac{a^2}{6} + \frac{b^2}{6} + \frac{c^2}{6} + \frac{2bc\sqrt{3}}{6}\sin\alpha.$$

In the end, using the relation $S = 1/2bc\sin\alpha$ for the area S of a triangle ABC, we obtain

$$|B'C'|^2 = \frac{a^2+b^2+c^2}{6} + \frac{2}{\sqrt{3}}S. \tag{6.4}$$

The side length $|B'C'|$ is expressed by the formula (6.4) which is symmetric in a, b, c. Then it is obvious that the same expression we get for the remaining sides $C'A'$ and $A'B'$. Thus a triangle $A'B'C'$ is equilateral and Napoleon's theorem is proved.

Proving Napoleon's theorem for the inner Napoleon triangle $A''B''C''$ we get analogously to (6.4) the formula

$$|B''C''|^2 = \frac{a^2+b^2+c^2}{6} - \frac{2}{\sqrt{3}}S, \tag{6.5}$$

which differs from (6.4) only by the sign.

If we denote S', S'' the areas of Napoleon's triangles $A'B'C'$ and $A''B''C''$ respectively, then in accordance with (6.4), (6.5),

$$S' = (a^2+b^2+c^2)\frac{\sqrt{3}}{24} + \frac{1}{2}S \tag{6.6}$$

and

$$S'' = (a^2+b^2+c^2)\frac{\sqrt{3}}{24} - \frac{1}{2}S \tag{6.7}$$

from which

$$S' - S'' = S$$

follows.

We proved that the sum of signed areas of triangles $A'B'C'$, $A''B''C''$ is equal to the sum of an original triangle ABC (considering the area S'' with the sign minus since the triangle $A''B''C''$ is oppositely oriented).

Remark 6.2. Since $|B''C''|^2 \geq 0$ in (6.5) we get from the formula (6.5) the sum of squares inequality for a triangle [15]:

Given a triangle with side lengths a, b, c and area S. Then

$$a^2 + b^2 + c^2 - 4\sqrt{3}\, S \geq 0. \tag{6.8}$$

The equality is attained only for an equilateral triangle.

The inequality (6.8) appeared in 1919 due to R. Weitzenböck [139, 40]. It can be considered as the type of isoperimetric inequality. In Chapter 7 we shall prove the inequality (6.8) by computer.

Another proof of Napoleon's theorem follows from the tessellation which is due to J. F. Rigby [110]. In the situation in Fig. 6.6 the centers of

Fig. 6.6 Napoleon's theorem — proof by tessellation

small equilateral triangles erected over sides of dark triangles form a regular triangular net which covers the whole plane. Midpoints of other equilateral triangles form the centers of this triangular net. In this way we get a new — finer — regular triangular net. From this, our statement follows.

6.1.1 Topics related to Napoleon's theorem

The next theorem is closely tied with Napoleon's theorem (Fig. 6.7):

Let A', B', C' be the vertices of equilateral triangles erected on the sides of

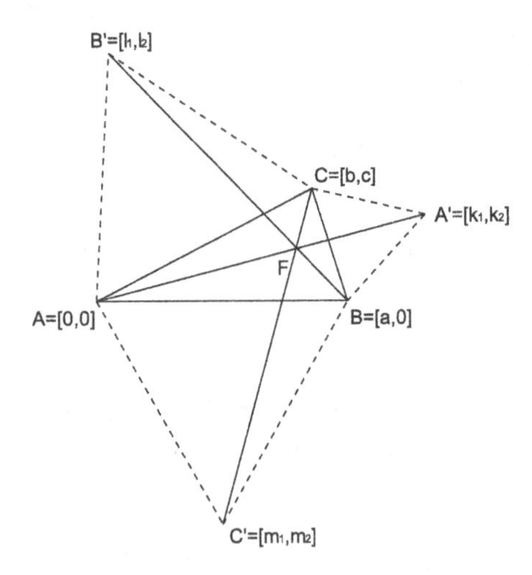

Fig. 6.7 The lines AA', BB', CC' are concurrent at a point F and $|AA'| = |BB'| = |CC'|$

a triangle ABC. Then the lines AA', BB', CC' are concurrent and $|AA'| = |BB'| = |CC'|$.

The fact that the lines AA', BB', CC' are concurrent at a point F will be shown later in this chapter in Section 6.1.2 "Kiepert hyperbola".

In order to prove the equality $|AA'| = |BB'| = |CC'|$ we will proceed similarly as by the proof of the Napoleon's theorem.

Recall that we denote $A' = [k_1, k_2]$, $B' = [l_1, l_2]$, $C' = [m_1, m_2]$. We will search for such a value v for which $|AA'| = |BB'|$ and $|BB'| = |CC'|$. We have (Fig. 6.7):

$$|AA'| = |BB'| \Leftrightarrow z_3 : k_1^2 + k_2^2 - (l_1 - a)^2 - l_2^2 = 0,$$

$$|BB'| = |CC'| \Leftrightarrow z_4 : (l_1 - a)^2 + l_2^2 - (m_1 - b)^2 - (m_2 - c)^2 = 0.$$

First we will explore the case $|AA'| = |BB'|$. In the ideal $I = (h_1, h_2, \ldots, h_6, z_3)$ we eliminate all variables up to a, b, c, v. We obtain

```
Use R::=Q[k[1..2]l[1..2]m[1..2]abcv];
```

```
I:=Ideal(2k[1]-a-b-2vc,2k[2]-c-2va+2vb,2l[1]-b+2vc,2l[2]-c-2vb,
2m[1]-a,m[2]+va,k[1]^2+k[2]^2-(l[1]-a)^2-l[2]^2);
Elim(k[1]..m[2],I);
```

the polynomial $a(a-2b)(4v^2-3)$ from which

a) $a-2b=0$, that is, $|BC|=|AC|$ and the equality $|AA'|=|BB'|$ holds for an arbitrary value v,

or

b) $4v^2-3=0$, that is, $v=\sqrt{3}/2$ or $v=-\sqrt{3}/2$ and an arbitrary triangle ABC (we rule out the case $a=0$).

For the value $v=\sqrt{3}/2$ we get equilateral triangles BCA' and ACB' erected outwardly of a triangle ABC, whereas for $v=-\sqrt{3}/2$ we get triangles BCA' and ACB' erected inwardly. These triangles are solutions of our problem. The verification

```
Use R::=Q[k[1..2]l[1..2]m[1..2]abcvt];
J:=Ideal(2k[1]-a-b-2vc,2k[2]-c-2va+2vb,2l[1]-b+2vc,2l[2]-c-2vb,
2m[1]-a,m[2]+va,4v^2-3,at-1);
NF(k[1]^2+k[2]^2-(l[1]-a)^2-l[2]^2,J);
0
```

confirms that $|AA'|=|BB'|$. In an analogous way we prove that $|BB'|=|CC'|$ also holds.

Now we prove that $|AA'|=|BB'|=|CC'|$ classically (Fig. 6.7). Consider the triangles $AC'C$ and ABB'. These triangles are congruent since $|AC'|=|AB|$, $|AC|=|AB'|$ and $|\angle CAC'|=|\angle B'AB|=\alpha+60^\circ$. Rotation by the angle $\alpha+60^\circ$ with the center A maps the triangle $AC'C$ into ABB'. This implies $|CC'|=|BB'|$. Similarly we proceed to prove that $|AA'|=|BB'|$.

The situation in Fig. 6.7 is often called the *Torricelli configuration* [73, 6]. The intersection of the lines AA', BB', CC' is a point F which is known as Fermat or Fermat–Torricelli point which has the following property:

If no angle of ABC exceeds 120° then the point F minimizes the sum of distances from an arbitrary point X in the plane of ABC to the vertices A, B, C.

We shall prove it classically by means of Napoleon's theorem [40] (Fig. 6.8). Let $A_1B_1C_1$ be the outer Napoleon triangle of ABC. It is obvious that the

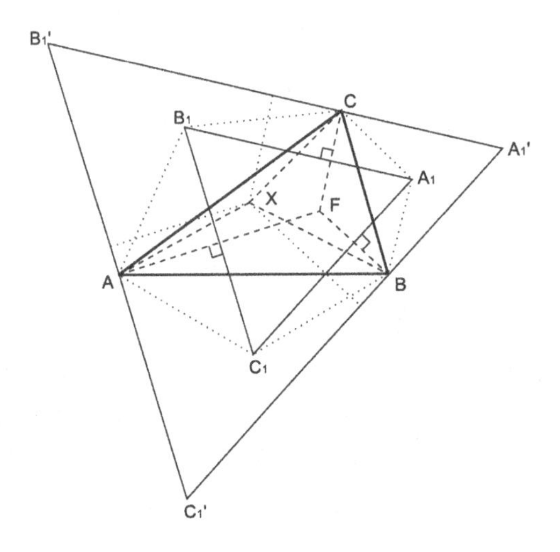

Fig. 6.8 The minimal property of the Fermat's point F: $|AX|+|BX|+|CX| \geq |AF|+|BF|+|CF|$

circumcircles of the triangles ABC_1, BCA_1 and CAB_1 are concurrent at F. This implies that A_1B_1, B_1C_1, C_1A_1 are perpendicular bisectors of CF, AF, BF respectively. Hence $|AF|+|BF|+|CF|$ equals two times the sum of distances of F to the sides of the equilateral Napoleon's triangle $A_1B_1C_1$ and this is equal to the doubled height of $A_1B_1C_1$ (see Viviani's theorem in Chapter 9). Let us construct through the vertices of ABC parallel lines to the sides B_1C_1, C_1A_1, A_1B_1. We obtain an equilateral triangle $A_1'B_1'C_1'$ whose sides are parallel to the Napoleon triangle $A_1B_1C_1$. From the previous considerations we get that $|AF|+|BF|+|CF|$ equals the height h of $A_1'B_1'C_1'$. Now X being an arbitrary point inside ABC we get that $|AX|+|BX|+|CX|$ is greater than or equal to the sum of distances of X to the sides $A_1'B_1'C_1'$ which is h.

Now we should expect a computer proof of the fact, that the point F has minimal property. Despite great effort of the author all attempts have failed. See Chapter 7 to read about the difficulties encountered.

Remark 6.3. The minimal sum of distances $|FA|+|FB|+|FC|$ is equal to the lengths $|CC'| = |AA'| = |BB'|$ as well. The point F is also the *isogonic* point of ABC, i.e., a point from which we see all the sides of a triangle

ABC under the same angle 120°.

6.1.2 *Kiepert hyperbola*

Further we will prove a theorem which is ascribed to L. Kiepert [61, 73, 140, 22, 110]. He used this theorem to solve the problem: *Construct a triangle ABC if the vertices A', B', C' of the outer Napoleon's triangle of ABC are given.* The theorem is as follows, cf. [136, 135]:

Theorem 6.2. *On the sides of a triangle ABC construct similar isosceles triangles ABC', BCA', CAB' (all outwardly or all inwardly). Then the lines AA', BB', CC' are concurrent at a point S.*

In a chosen Cartesian system of coordinates we have $A = [0,0]$, $B = [a,0]$, $C = [b,c]$, $A' = [k_1,k_2]$, $B' = [l_1,l_2]$, $C' = [m_1,m_2]$ (Fig. 6.9). We will pro-

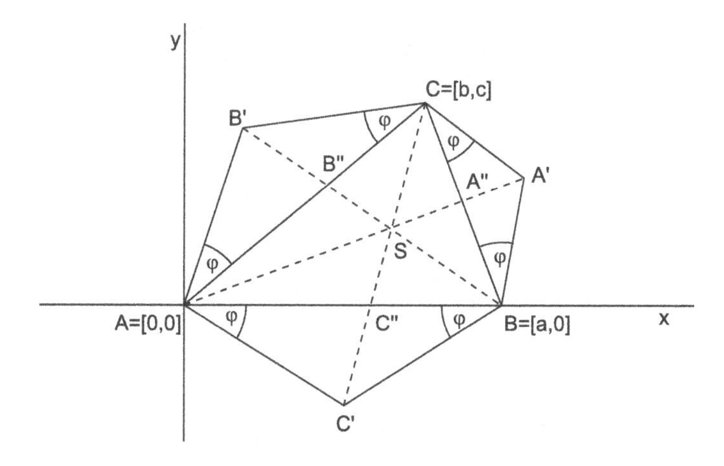

Fig. 6.9 Lines AA', BB', CC' intersect at the point S

ceed similarly as in the proof of Napoleon's theorem. From the construction of the points A', B', C' (see the proof of the theorem of Napoleon) the relations

$h_1 : 2k_1 - a - b - 2vc = 0,$
$h_2 : 2k_2 - c - 2va + 2vb = 0,$
$h_3 : 2l_1 - b + 2vc = 0,$
$h_4 : 2l_2 - c - 2vb = 0,$
$h_5 : 2m_1 - a = 0,$

$h_6 : m_2 + va = 0$

follow. The number v is an arbitrary real number which determines the form and orientation of the similar triangles ABC', BCA', CAB'. Further suppose that $S = [s_1, s_2]$ is a common point of the straight lines AA' and BB'. We are to prove that the point S is on the line CC' as well. We have

$S \in AA' \Leftrightarrow h_7 : k_1 s_2 - k_2 s_1 = 0,$
$S \in BB' \Leftrightarrow h_8 : (l_1 - a)s_2 - (s_1 - a)l_2 = 0.$

The conclusion z is of the form

$S \in CC' \Leftrightarrow z : (b - m_1)(s_2 - m_2) - (s_1 - m_1)(c - m_2) = 0.$

We want to prove that the conclusion polynomial z belongs to the ideal $I = (h_1, h_2, \ldots, h_8)$. We enter

```
Use R::=Q[k[1..2]l[1..2]m[1..2]s[1..2]abcv];
I:=Ideal(2k[1]-a-b-2vc,2k[2]-c-2va+2vb,2l[1]-b+2vc,2l[2]-c-2vb,
2m[1]-a,m[2]+va,k[1]s[2]-k[2]s[1],(l[1]-a)s[2]-(s[1]-a)l[2]);
NF((b-m[1])(s[2]-m[2])-(s[1]-m[1])(c-m[2]),I);
```

and get the answer 0 which means that the lines AA', BB', CC' are concurrent at the point S.

Now a classical proof of the statement above, which was published by O. Bottema [17], follows. This proof is short and elegant and we think that it is worth reproducing. It is based on the area principle which we know from Chapter 5. By Fig. 6.9 it holds:

$$|AC''|/|C''B| = |ACC'|/|BCC'|$$

$$= |AC||AC'|\sin(A + \varphi)/|BC||BC'|\sin(B + \varphi)$$

$$= |AC|\sin(A + \varphi)/|BC|\sin(B + \varphi),$$

where $|ACC'|$ denotes the area of ACC', etc. Similarly,

$$|BA''|/|A''C| = |AB|\sin(B + \varphi)/|AC|\sin(C + \varphi),$$

$$|CB''|/|B''A| = |BC|\sin(C + \varphi)/|AB|\sin(A + \varphi).$$

We arrive at the equality

$$\frac{|AC''|}{|C''B|}\frac{|BA''|}{|A''C|}\frac{|CB''|}{|B''A|} = 1,$$

from which our statement from the converse of Ceva's theorem follows.

Remark 6.4. The previous theorem has a wide application. For various values of the base angle φ we get various similar isosceles triangles which correspond to various points S.

For the value $\varphi = 60°$ we get equilateral triangles ABC', BCA', CAB', erected *outwardly* of a triangle ABC which leads to the Fermat point F (Fig. 6.7). For $\varphi = 0°$ the vertices of isosceles triangles form midpoints of the sides of a triangle ABC and their connecting lines with opposite vertices intersect at the centroid of ABC. For the value $-60°$ we get the second Fermat point F', the angle $\varphi = 90°$ gives the orthocenter of ABC, etc.

We proved that the lines AA', BB', CC' are concurrent at a point S which indeed varies in accordance with the change of similar isosceles triangles ABC', BCA', CAB' constructed on the sides of a triangle ABC (Fig. 6.9). Let us show the locus of points S by varying the value v (and the base angle φ).

With the same notation as in the last example we eliminate variables $k_1, k_2, l_1, l_2, m_1, m_2, v$ in the ideal $I = (h_1, h_2, \ldots, h_8)$. Then the elimination ideal contains only the polynomials in the variables a, b, c, s_1, s_2 (Fig. 6.10). We get

```
Use R::=Q[k[1..2]l[1..2]m[1..2]vs[1..2]abc];
I:=Ideal(2k[1]-a-b-2vc,2k[2]-c-2va+2vb,2l[1]-b+2vc,2l[2]-c-2vb,
2m[1]-a,m[2]+va,k[1]s[2]-k[2]s[1],(l[1]-a)s[2]-(s[1]-a)l[2]);
Elim(k[1]..v,I);
```

the polynomial which gives the equation of a conic (where we write $[x, y]$ instead $[s_1, s_2]$)

$$x^2c(a-2b)+2xy(a^2-ab+b^2-c^2)+y^2c(2b-a)+xac(2b-a)+ya(c^2-ab-b^2) = 0, \tag{6.9}$$

that is called the *Kiepert hyperbola* [61, 73]. Hence the point S lies on the hyperbola (6.9). Do the points S fill in the whole hyperbola or only part of it? To show this we compute the value v for an arbitrary point $S = [x, y]$ which obeys (6.9). After a short calculation we find out that for values $x = b,\ y = (a - b)b/c$, which correspond to the orthocenter of a triangle, the value v does not exist. Thus the locus of the points S is the hyperbola (6.9) without the orthocenter of a triangle ABC.

The Kiepert hyperbola (6.9) has many interesting properties which can be obtained both classically and from the equation (6.9). It is a rectangular hyperbola which passes through the vertices of a triangle ABC. It contains

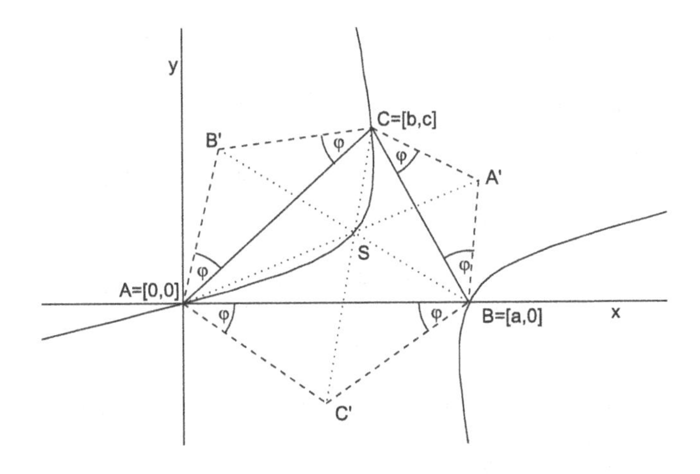

Fig. 6.10 The point S lies on the Kiepert hyperbola

other "remarkable" points of ABC such as the centroid, the orthocenter, the outer and inner Fermat's points, etc. The Kiepert hyperbola is closely tied with the Simson–Wallace line, the Feuerbach circle, etc. (see [73, 108]).

6.1.3 *Generalization of Napoleon's theorem*

Napoleon's theorem can be generalized in several ways. One of them is as follows [26, 73, 109]:

Theorem 6.3. *If three similar triangles $A'BC$, $AB'C$ ABC' are erected (all outwardly or all inwardly) on the sides of an arbitrary triangle ABC, then their circumcenters R, S, T form a triangle similar to the three triangles.*

First we will prove the statement by computer.

Choose a Cartesian system of coordinates so that $A = [0,0]$, $B = [1,0]$, $C = [p,q]$, $A' = [a_1, a_2]$, $B' = [b_1, b_2]$, $C' = [c_1, c_2]$, $R = [r_1, r_2]$, $S = [s_1, s_2]$, $T = [t_1, t_2]$ (Fig. 6.11). We are to describe similar triangles $A'BC$, $AB'C$ ABC' erected on the sides of ABC all outwardly or all inwardly analytically. We will explore the outward case.

Projecting orthogonally the vertex A' onto the side BC we obtain a point K. Analogously we project the point A onto the sides $B'C$ and BC' to get the points L and M. It is obvious that the ratios $(BCK), (B'CL), (BC'M)$ of the points K, L, M with respect to the end-

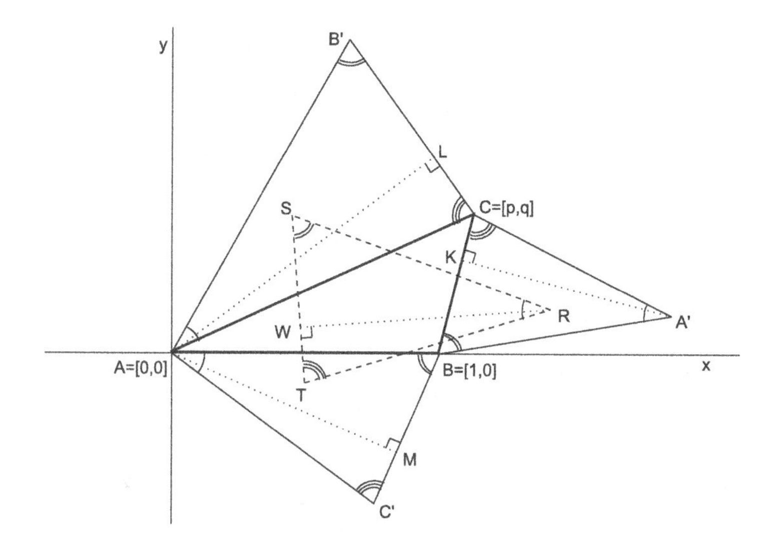

Fig. 6.11 Triangle RST is similar to three triangles $A'BC, AB'C, ABC'$

points of the sides $BC, B'C, BC'$ of similar triangles $A'BC, AB'C, ABC'$ are equal. Denote this common value of the ratios by u. The vertex A' is the endpoint of a vector whose initial point is at the point K with the length $v|KB|$, where v is a real number, and the same direction as the vector $B-K$ rotated by the angle 90° in a positive sense. We denote it as $A'-K = v \cdot \mathrm{rot}_{+90^\circ}(B-K)$, where $\mathrm{rot}_{+90^\circ}(B-K)$ is the vector $B-K$ rotated by 90° in a positive sense. Similarly we express the respective vectors $A-L$ and $A-M$ and get the relations $A-L = v \cdot \mathrm{rot}_{+90^\circ}(B'-L)$ and $A-M = v \cdot \mathrm{rot}_{+90^\circ}(B-M)$. The real numbers u, v are the same for all points K, L, M because of similarity of the triangles $A'BC$, $AB'C$, ABC'. We can say that the parameters u, v characterize these similar triangles.

To prove that RST is similar to $A'BC$, $AB'C$, ABC' it suffices to show that the parameters u, v obey the relation $R-W = v \cdot \mathrm{rot}_{-90^\circ}(S-W)$, where W fulfils $(STW) = u$. Here $\mathrm{rot}_{-90^\circ}(S-W)$ is the vector $S-W$ rotated by the angle 90° in a negative sense.

Let us denote the coordinates of the remaining points K, L, M, W by $K = [k_1, k_2]$, $L = [l_1, l_2]$, $M = [m_1, m_2]$, $W = [w_1, w_2]$. Now we can translate the situation into algebraic form:

$(BCK) = u \Leftrightarrow$
$h_1 : k_1 - 1 - u(k_1 - p) = 0,\ h_2 : k_2 - u(k_2 - q) = 0,$

$(B'CL) = u \Leftrightarrow$
$h_3 : l_1 - b_1 - u(l_1 - p) = 0,\ h_4 : l_2 - b_2 - u(l_2 - q) = 0,$

$(BC'M) = u \Leftrightarrow$
$h_5 : m_1 - 1 - u(m_1 - c_1) = 0,\ h_6 : m_2 - u(m_2 - c_2) = 0,$

$A' - K = v \cdot \mathrm{rot}_{+90^\circ}(B - K) \Leftrightarrow$
$h_7 : a_1 - k_1 - vk_2 = 0,\ h_8 : a_2 - k_2 - v(1 - k_1) = 0,$

$A - L = v \cdot \mathrm{rot}_{+90^\circ}(B' - L) \Leftrightarrow$
$h_9 : -l_1 - v(l_2 - b_2) = 0,\ h_{10} : -l_2 - v(b_1 - l_1) = 0,$

$A - M = v \cdot \mathrm{rot}_{+90^\circ}(B - M) \Leftrightarrow$
$h_{11} : -m_1 - vm_2 = 0,\ h_{12} : -m_2 - v(1 - m_1) = 0,$

R is the circumcenter of $A'BC \Leftrightarrow$
$h_{13} : 2(p-1)r_1+2qr_2+1-p^2-q^2 = 0,\ h_{14} : 2(a_1-1)+2a_2r_2+1-a_1^2-a_2^2 = 0,$

S is the circumcenter of $AB'C \Leftrightarrow$
$h_{15} : 2ps_1 + 2qs_2 - p^2 - q^2 = 0,\ h_{16} : 2b_1s_1 + 2b_2s_2 - b_1^2 - b_2^2 = 0,$

T is the circumcenter of $ABC' \Leftrightarrow$
$h_{17} : 2t_1 - 1 = 0,\ h_{18} : 2c_1t_1 + 2c_2t_2 - c_1^2 - c_2^2 = 0,$

$(STW) = u \Leftrightarrow h_{19} : w_1 - s_1 - u(w_1 - t_1) = 0,\ h_{20} : w_2 - s_2 - u(w_2 - t_2) = 0.$

The conclusion polynomials z_1, z_2 are of the form

$R - W = v \cdot \mathrm{rot}_{-90^\circ}(S - W) \Leftrightarrow$
$z_1 : r_1 - w_1 - v(s_2 - w_2) = 0,\ z_2 : r_2 - w_2 - v(w_1 - s_1) = 0.$

Consider the ideal $I = (h_1, h_2, \ldots, h_{20})$. First explore the conclusion polynomial z_1.

The normal form of 1 in the ideal $J = I \cup \{z_1x - 1\}$, where x is a slack variable, is equal to 1 which implies that z_1 does not belong to the radical $\sqrt{I}$.

Searching for subsidiary conditions does not give in CoCoA any result because of the complexity of the system of algebraic equations. Therefore we try "to guess" unknown additional conditions. Adding the polynomial $u + 1$ to the ideal J, i.e. considering that $u = -1$, and eliminating all variables except p, q we get

```
Use R::=Q[pquvk[1..2]l[1..2]m[1..2]a[1..2]b[1..2]c[1..2]
s[1..2]t[1..2]w[1..2]r[1..2]x];
K:=Ideal(k[1]-1-u(k[1]-p),k[2]-u(k[2]-q),l[1]-b[1]-u(l[1]-p),
```

```
l[2]-b[2]-u(l[2]-q),m[1]-1-u(m[1]-c[1]),m[2]-u(m[2]-c[2]),a[1]
-k[1]-vk[2],a[2]-k[2]-v(1-k[1]),-l[1]-v(l[2]-b[2]),-l[2]-v(b[1]
-l[1]),-m[1]-vm[2],-m[2]-v(1-m[1]),2(p-1)r[1]+2qr[2]+1-p^2-q^2,
2(a[1]-1)r[1]+2a[2]r[2]+1-a[1]^2-a[2]^2,2ps[1]+2qs[2]-p^2-q^2,
2b[1]s[1]+2b[2]s[2]-b[1]^2-b[2]^2,2t[1]-1,2c[1]t[1]+2c[2]t[2]-
c[1]^2-c[2]^2,w[1]-s[1]-u(w[1]-t[1]),w[2]-s[2]-u(w[2]-t[2]),
u+1,(r[1]-w[1]-v(s[2]-w[2]))x-1);
Elim(u..x,K);
```

the condition $(p^2+q^2)((p-1)^2+q^2) = 0$ which means that $C = A$ or $C = B$. We obtain the same condition by adding polynomials of the form $u-k$ or $v-l$ for various real constants k, l to the ideal J. This leads us to the idea to add to the ideal J a subsidiary condition given by the polynomial $(p^2+q^2)((p-1)^2+q^2)y - 1$ which means that $C \neq A$, $C \neq B$, where y is a slack variable. We enter

```
Use R::=Q[pquvk[1..2]l[1..2]m[1..2]a[1..2]b[1..2]c[1..2]s[1..2]
t[1..2]w[1..2]r[1..2]xy];
L:=Ideal(k[1]-1-u(k[1]-p),k[2]-u(k[2]-q),l[1]-b[1]-u(l[1]-p),
l[2]-b[2]-u(l[2]-q),m[1]-1-u(m[1]-c[1]),m[2]-u(m[2]-c[2]),a[1]
-k[1]-vk[2],a[2]-k[2]-v(1-k[1]),-l[1]-v(l[2]-b[2]),-l[2]-v(b[1]
-l[1]),-m[1]-vm[2],-m[2]-v(1-m[1]),2(p-1)r[1]+2qr[2]+1-p^2-q^2,
2(a[1]-1)r[1]+2a[2]r[2]+1-a[1]^2-a[2]^2,2ps[1]+2qs[2]-p^2-q^2,
2b[1]s[1]+2b[2]s[2]-b[1]^2-b[2]^2,2t[1]-1,2c[1]t[1]+2c[2]t[2]-
c[1]^2-c[2]^2,w[1]-s[1]-u(w[1]-t[1]),w[2]-s[2]-u(w[2]-t[2]),
(p^2+q^2)((p-1)^2+q^2)y-1,(r[1]-w[1]-v(s[2]-w[2]))x-1);
NF(1,L);
```

and CoCoA returns 0.
We get a similar result for the second conclusion polynomial z_2. Theorem 6.3 is proved.

Now we shall give a classical proof of the above theorem which is due to D. Pedoe [94]. We will proceed similarly as by the classical proof of Napoleon's theorem 6.1.

Let ABC and KLM be arbitrary triangles and a, b, c and k, l, m their side lengths respectively. On the sides of ABC outwardly construct triangles $A'BC$, $AB'C$ and ABC' which are similar to KLM. Let R, S, T be the circumcenters and r_a, r_b, r_c the circumradii of $A'BC$, $AB'C$, ABC' respectively (Fig. 6.12). Then

$$\frac{a}{k} = \frac{b}{l} = \frac{c}{m} = \frac{r_a}{r} = \frac{r_b}{r} = \frac{r_c}{r},$$

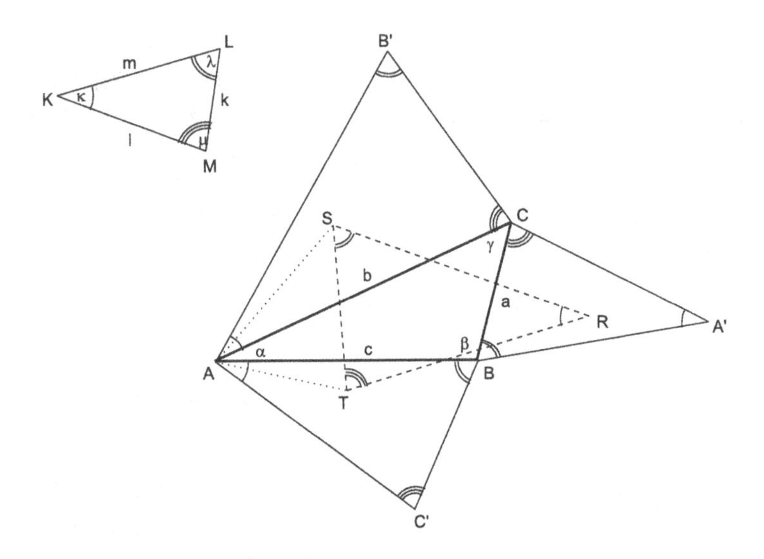

Fig. 6.12 Generalized Napoleon's theorem — classical proof

where r is the circumradius of KLM.

By the law of cosines applied to the triangle AST we get

$$|ST|^2 = |AS|^2 + |AT|^2 - 2|AS||AT|\cos(\alpha+\kappa) = r_b^2 + r_c^2 - 2r_b r_c \cos(\alpha+\kappa),$$

$$|ST|^2 = r^2\left(\frac{b^2}{l^2} + \frac{c^2}{m^2} - 2\,\frac{b}{l}\frac{c}{m}(\cos\alpha\cos\kappa - \sin\alpha\sin\kappa)\right).$$

Further we get

$$|ST|^2 = r^2\left(\frac{b^2}{l^2} + \frac{c^2}{m^2} - 2\,\frac{bclm}{l^2m^2}\,(\cos\alpha\cos\kappa - \sin\alpha\sin\kappa)\right),$$

$$2|ST|^2 = \frac{r^2}{l^2m^2}\,(2m^2b^2 + 2l^2c^2 - 2bc\cos\alpha\;2lm\cos\kappa + 2bc\sin\alpha\;2lm\sin\kappa).$$

Since

$2bc\cos\alpha = b^2 + c^2 - a^2,\; 2lm\cos\kappa = l^2 + m^2 - k^2,$

$bc\sin\alpha = 2p,\; lm\sin\kappa = 2q,$

where p and q are areas of the triangles ABC and KLM, then

$$2|ST|^2 = \frac{r^2}{l^2m^2}(2m^2b^2 + 2l^2c^2 - (b^2+c^2-a^2)(l^2+m^2-k^2) + 16pq),$$

which becomes

$$2\left(\frac{|ST|}{k}\right)^2 = \frac{r^2}{k^2l^2m^2}(k^2(-a^2+b^2+c^2)+l^2(a^2-b^2+c^2)+m^2(a^2+b^2-c^2)+16pq). \tag{6.10}$$

We see that on the right-hand side of (6.10) no variable a, b, c, k, l, m is preferred. Thus for other side lengths $|TR|/l$ and $|RS|/m$ we obtain the same result. From this we get

$$\frac{|RS|}{m} = \frac{|ST|}{k} = \frac{|TR|}{l}$$

which means that the triangle RST is similar to the triangle KLM. Theorem 6.3 is proved.

Remark 6.5. If a triangle KLM is equilateral then Theorem 6.3 becomes Napoleon's theorem 6.1.

The classical proof of Theorem 6.3 above can be used to prove classically the Neuberg–Pedoe inequality (Theorem 7.8) from Chapter 7, where the Neuberg–Pedoe inequality is proved automatically by computer. It is as follows [94]:

Given two triangles ABC and KLM with the areas p, q and side lengths a, b, c and k, l, m respectively. Then

$$k^2(-a^2+b^2+c^2)+l^2(a^2-b^2+c^2)+m^2(a^2+b^2-c^2) \geq 16pq. \tag{6.11}$$

The equality is attained if and only if the triangles ABC and KLM are similar.

Consider similar triangles $A'BC, AB'C, ABC'$ which are erected *inwardly* on the sides of a triangle ABC. With the same notation as in the classical proof above we may write

$$|ST|^2 = |AS|^2 + |AT|^2 - 2|AS||AT|\cos(\alpha-\kappa) = r_b^2 + r_c^2 - 2r_br_c\cos(\alpha-\kappa),$$

which can be written as

$$2\left(\frac{|ST|}{k}\right)^2 = \frac{r^2}{k^2l^2m^2}(k^2(-a^2+b^2+c^2)+l^2(a^2-b^2+c^2)+m^2(a^2+b^2-c^2)-16pq). \tag{6.12}$$

From (6.12) we see that the right-hand side is greater than or equal to zero since the left side is non-negative.

The equality in (6.12) is attained if and only if $|ST| = |TR| = |RS| = 0$, which is fulfilled only in the case $\alpha = \kappa,\ \beta = \lambda,\ \gamma = \mu$, i.e, if the triangles ABC and KLM are similar.

In addition, from (6.12) we get that the triangle RST is similar to the three triangles $A'BC, AB'C, ABC'$.

Remark 6.6. J. F. Rigby [109] calls the triangle RST in Theorem 6.3 anti-similar to the triangles $A'BC$, $AB'C$ ABC' since RST has opposite orientation (is indirectly similar).

6.2 PDN theorem for a quadrilateral

In this part we will show a few special cases of the PDN theorem for a quadrilateral. First we prove the theorem of Finney. Then we will give the PDN theorem for a parallelogram which is also known as the Thébault's theorem. In the next part we will be concerned with a theorem which was published by M. H. van Aubel. Next some related theorems are given — the Napoleon–Barlotti theorem, the theorem of Finsler–Hadwiger, etc. Then the PDN theorem for a quadrilateral is given. This section is concluded by the PDN theorem in its general form for an arbitrary n-gon.

6.2.1 *Theorem of Finney*

One problem connected with the PDN theorem is a theorem which was published by R. L. Finney [39]:

Theorem 6.4 (Finney). *Let A', B' be centers of squares erected on the sides BC, AC of a triangle ABC (both outwardly or both inwardly) and let P be the midpoint of the side AB. Then $A'B'P$ is a right triangle for which $A'P \perp B'P$ and $|A'P| = |B'P|$.*

First we will "discover" this theorem.

Choose a Cartesian coordinate system so that $A = [0, 0]$, $B = [a, 0]$, $C = [b, c]$. Further denote $A' = [k_1, k_2]$, $B' = [l_1, l_2]$, $P = [p, 0]$ (Fig. 6.13). Instead of right isosceles triangles we will construct on the sides AC, BC as the bases of a triangle ABC *arbitrary* similar isosceles triangles ACB', CBA'. The point B' is the endpoint of a vector whose initial point is in the center of AC with the length $v|AC|$, where v is an arbitrary number, and the same direction as the vector $C - A$ rotated by the angle 90° in a positive sense. For the coordinates l_1, l_2 of the point B' the equality $(l_1 - b/2, l_2 - c/2) = v(-c, b)$ holds. Similarly we define the point A' for which $(k_1 - (a + b)/2, k_2 - c/2) = v(c, a - b)$. We have:

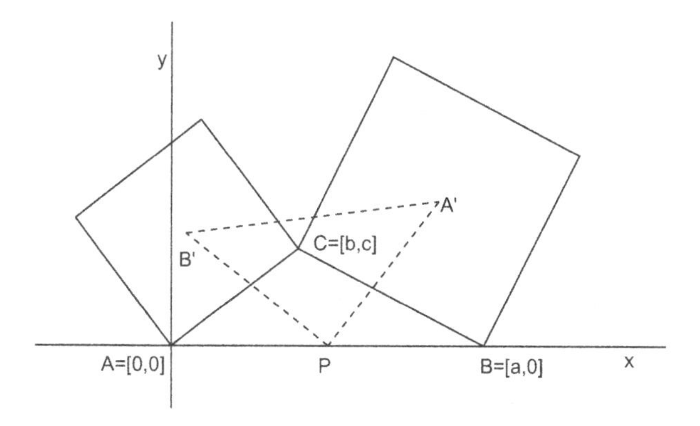

Fig. 6.13 Finney's theorem — $A'B'P$ is a right isosceles triangle

$A' - (B+C)/2 = v \cdot \mathrm{rot}(B-C) \Leftrightarrow$
$h_1 : 2k_1 - a - b - 2vc = 0,\ h_2 : 2k_2 - c - 2va + 2vb = 0,$

$B' - (A+C)/2 = v \cdot \mathrm{rot}(C-A) \Leftrightarrow$
$h_3 : 2l_1 - b + 2vc = 0,\ h_4 : 2l_2 - c - 2vb = 0,$

P is midpoint of $AB \Leftrightarrow h_5 : 2p - a = 0,$

$A'P \perp B'P \Leftrightarrow h_6 : (k_1 - p)(l_1 - p) + k_2 l_2 = 0,$

$|A'P| = |B'P| \Leftrightarrow h_7 : (k_1 - p)^2 + k_2^2 - (l_1 - p)^2 - l_2^2 = 0.$

First we shall search for such a real v that the condition $A'P \perp B'P$ is fulfilled. Elimination of variables k_1, k_2, l_1, l_2, p in the ideal $I = (h_1, h_2, \ldots, h_6)$

```
Use R::=Q[k[1..2]l[1..2]pabcv];
I:=Ideal(2k[1]-a-b-2vc,2k[2]-c-2va+2vb,2l[1]-b+2vc,2l[2]-c-2vb,
2p-a,(k[1]-p)(l[1]-p)+k[2]l[2]);
Elim(k[1]..p,I);
```

returns the equation $(2v+1)(2v-1)(ab-b^2-c^2) = 0$ which gives conditions $v = 1/2$ or $v = -1/2$ or the condition $ab - b^2 - c^2 = 0$ which means that $AC \perp BC$, that is, ABC is a right triangle.

Now we will explore the second condition $|A'P| = |B'P|$. Replacing the polynomial h_6 in the ideal I by the polynomial h_7, we get the equation $(2v+1)(2v-1)a(a-2b) = 0$ which gives the values $v = 1/2$ or $v = -1/2$ or $a = 2b$, i.e. $|AC| = |BC|$, or $a = 0$, when a triangle ABC degenerates. We will rule out the last condition assuming that $a \neq 0$, i.e., $A \neq B$. Two possibilities may occur:

a) for a right isosceles triangle ABC with $|AC| = |BC|$, $AC \perp BC$, the triangle $A'B'P$ has the required properties *independent* of v,

b) for an *arbitrary* triangle ABC only values $v = 1/2$ or $v = -1/2$ come into consideration.

We will verify this fact by adding the polynomial $4v^2 - 1$ into the ideal $(h_1, h_2, \dots, h_5)$. We compute the normal form of the polynomial h_6 with respect to the Gröbner basis of the ideal $J = (h_1, h_2, \dots, h_5, 4v^2 - 1)$.

```
Use R::=Q[k[1..2]l[1..2]pabcv];
J:=Ideal(2k[1]-a-b-2vc,2k[2]-c-2va+2vb,2l[1]-b+2vc,2l[2]-c-2vb,
2p-a,4v^2-1);
NF((k[1]-p)^2+k[2]^2-(l[1]-p)^2-l[2]^2,J);
```

We get the answer `NF=0`.
Analogously we get `NF((k[1]-p)(l[1]-p)+k[2]l[2],J)=0`. The theorem of Finney is proved.

A classical proof of Finney's theorem follows from the classical proof of the theorem of Thébault in the next section.

6.2.2 *Thébault's theorem*

The following theorem was published in 1937 by V. Thébault [131] (Fig. 6.14):

Theorem 6.5 (Thébault). *On the sides of a parallelogram squares are erected (all outside or all inside). Then the centers of squares form a square.*

A computer proof of Thébault's theorem follows from the computer proof of Finney's theorem. Finney's theorem is equivalent to Thébault's theorem. This is obvious when we complete a triangle ABC into a parallelogram $ABCD$ (see Fig. 6.13).

The theorem of Thébault can be proved classically in the following way (the outer case) (Fig. 6.14).

The triangles $A'B'B$ and $C'B'C$ are congruent since $|B'B| = |B'C|$, $|BA'| = |CC'|$ and $|\angle B'BA'| = |\angle B'CC'|$. Rotation by the angle $90°$ with the center at the point B' maps the triangle $A'B'B$ into the triangle $C'B'C$. Whence $|A'B'| = |C'B'|$ and $A'B' \perp C'B'$. Similarly for other sides.

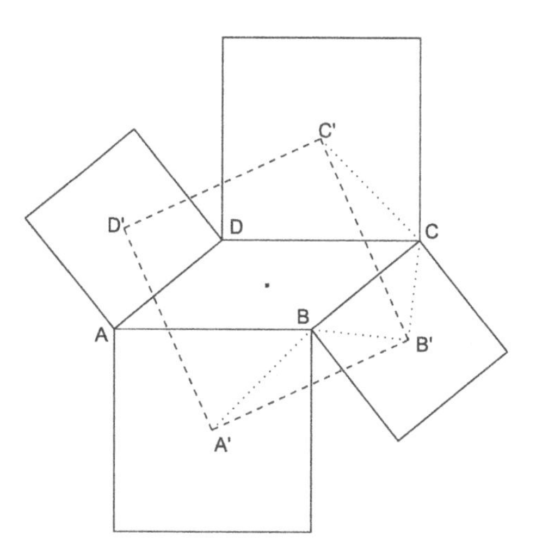

Fig. 6.14 Thébault's theorem — points A', B', C', D' form a square

An analogous proof can be applied if we construct squares inside a parallelogram. The same method can be used to prove the Finney's theorem classically.

The Thébault's theorem can also be proved by the following consideration. In Fig. 6.15 we can see a configuration from Fig. 6.14 which was used to perform a tessellation. Centers of small squares constructed over the

Fig. 6.15 PDN theorem for a quadrilateral — proof by a tessellation

sides of parallelograms form a square net, which covers the whole plane. Centers of big squares form the centers of this square net. We get a new — finer square net. From this our statement follows.

Remember that Thébault's theorem is both a special case of the PDN theorem [97], which will be given later, and a special case of the so-called theorem of Napoleon–Barlotti [73, 3] for $n = 4$:

Theorem 6.6 (Napoleon–Barlotti). *On the sides of an affine-regular n-gon construct regular n-gons (all outside or all inside). Then centers of these regular n-gons form a regular n-gon.*

Under affine-regular n-gon we understand an affine image of a regular n-gon (or simply spoken — its parallel projection). Hence an affine image of a regular (equilateral) triangle is an arbitrary triangle, whereas an affine image of a square is a parallelogram. These are special instances of the Napoleon–

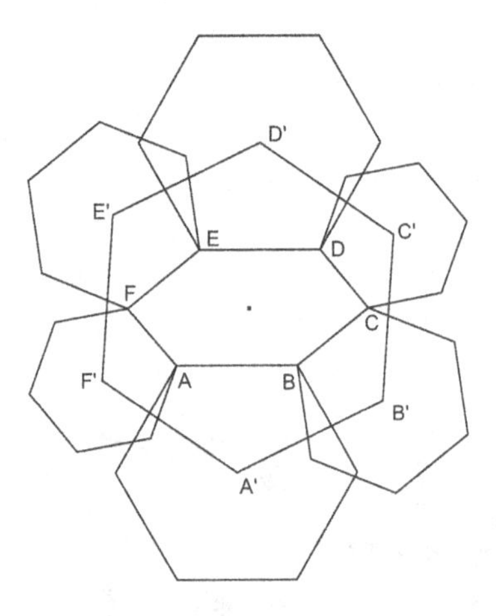

Fig. 6.16 Theorem of Napoleon–Barlotti — points A', B', C', D', E', F' form a regular hexagon

Barlotti theorem. In Fig. 6.16 the case for $n = 6$ of the Napoleon–Barlotti theorem is depicted.

6.2.3 *Theorem of Van Aubel*

Now we will give a few special cases of the PDN theorem. One of them gives the following theorem which was published in 1878 by M. H. van Aubel [1]:

Theorem 6.7 (Van Aubel). *On the sides of a quadrilateral $ABCD$ isosceles right triangles are erected. Their vertices form a quadrilateral $A'B'C'D'$ which fulfils*

$$A'C' \perp B'D', \quad |A'C'| = |B'D'|. \tag{6.13}$$

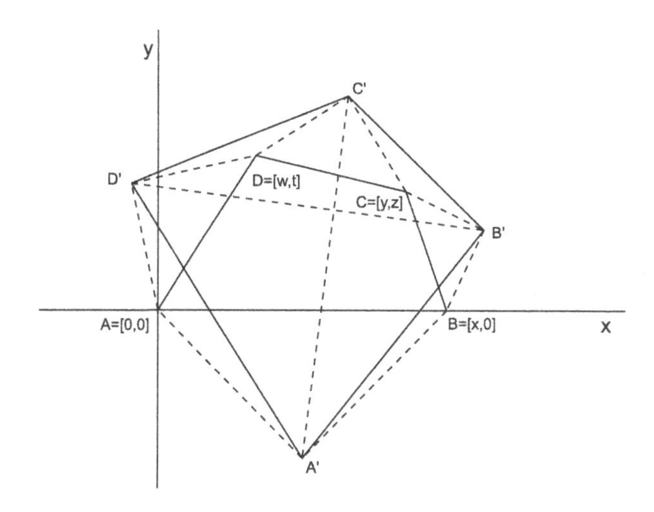

Fig. 6.17 Theorem of Van Aubel: $A'C' \perp B'D'$, $|A'C'| = |B'D'|$ — computer proof

We will prove (6.13) by computer.

On the sides of a quadrilateral $ABCD$ we construct isosceles right triangles ABA', BCB', CDD', DAD'. Choose a Cartesian coordinate system such that $A = [0,0]$, $B = [x,0]$, $C = [y,z]$, $D = [w,t]$ and let $A' = [a_1, a_2]$, $B' = [b_1, b_2]$, $C' = [c_1, c_2]$, $D' = [d_1, d_2]$ (Fig. 6.17). A point B' is the end-point of a vector whose initial point is in the center of BC with the length $1/2|BC|$, and the same direction as the vector $B - C$ rotated by the angle $90°$ in a positive sense. Then $(b_1 - (x+y)/2, b_2 - z/2) = 1/2(z, x-y)$. Analogously we define the other vertices A', C', D'. It holds

$A' - (A+B)/2 = 1/2 \text{ rot}(A - B) \Leftrightarrow$
$h_1 : 2a_1 - x = 0, \; h_2 : 2a_2 + x = 0,$

$B' - (B+C)/2 = 1/2 \text{ rot}(B - C) \Leftrightarrow$
$h_3 : 2b_1 - x - y - z = 0, \; h_4 : 2b_2 - z - x + y = 0,$

$C' - (C+D)/2 = 1/2 \text{ rot}(C - D) \Leftrightarrow$
$h_5 : 2c_1 - y - w - t + z = 0, \; h_6 : 2c_2 - z - t - y + w = 0,$

$D' - (D + A)/2 = 1/2 \operatorname{rot}(D - A) \Leftrightarrow$
$h_7 : 2d_1 - w + t = 0, \; h_8 : 2d_2 - t - w = 0.$

An algebraic translation of $|A'C'| = |B'D'|$ and $A'C' \perp B'D$ is as follows:

$|A'C'| = |B'D'| \Leftrightarrow$
$h_9 : (c_1 - a_1)^2 + (c_2 - a_2)^2 - (d_1 - b_1)^2 - (d_2 - b_2)^2 = 0,$

$A'C' \perp B'D' \Leftrightarrow$
$h_{10} : (c_1 - a_1)(d_1 - b_1) + (c_2 - a_2)(d_2 - b_2) = 0.$

First we shall prove that from the given assumptions $h_1, h_2, \ldots, h_8$ the condition h_9 follows. We ask whether the polynomial h_9 belongs to the ideal $I = (h_1, h_2, \ldots, h_8)$. We enter

```
Use R::=Q[a[1..2]b[1..2]c[1..2]d[1..2]xyzwt];
I:=Ideal(2a[1]-x,2a[2]+x,2b[1]-x-y-z,2b[2]-z-x+y,2c[1]-y-w-t+z,
2c[2]-z-t-y+w,2d[1]-w+t,2d[2]-t-w);
NF((c[1]-a[1])^2+(c[2]-a[2])^2-(d[1]-b[1])^2-(d[2]-b[2])^2,I);
```

and get the result NF=0 which means that $|A'C'| = |B'D'|$.

Exploring the condition $A'C' \perp B'D'$ we ask whether h_{10} is an element

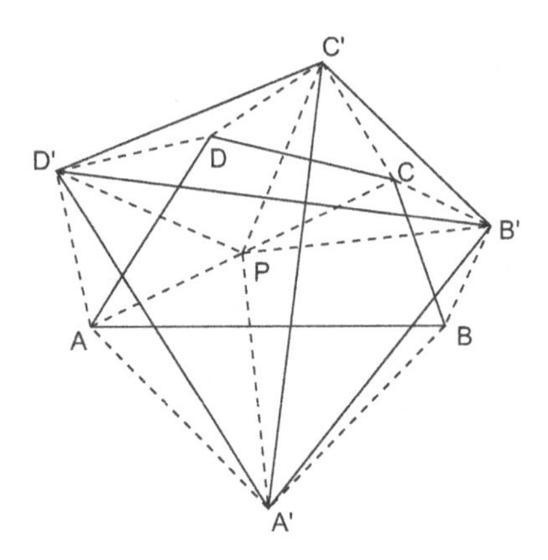

Fig. 6.18 Theorem of Van Aubel — classical proof

of the ideal $I = (h_1, h_2, \ldots, h_8)$.

```
Use R::=Q[a[1..2]b[1..2]c[1..2]d[1..2]xyzwt];
I:=Ideal(2a[1]-x,2a[2]+x,2b[1]-x-y-z,2b[2]-z-x+y,2c[1]-y-w-t+z,
2c[2]-z-t-y+w,2d[1]-w+t,2d[2]-t-w);
```

```
NF((c[1]-a[1])(d[1]-b[1])+(c[2]-a[2])(d[2]-b[2]),I);
```

CoCoA returns `NF=0`. Hence the conditions (6.13) are fulfilled and the theorem of Van Aubel is proved.

Now we give a classical proof of the theorem of Van Aubel. We apply the theorem of Finney to the triangles ABC and ACD. By this theorem $A'PB', C'PD'$ are isosceles right triangles, where P is the midpoint of the diagonal AC (Fig. 6.18). The triangles $A'PC'$ and $B'PD'$ are obviously congruent. Rotation with the center at P by the angle $90°$ maps the triangle $A'PC'$ into $B'PD'$. Whence $|A'C'| = |B'D'|$ and $A'C' \perp B'D'$.

The next theorem is connected with the previous one and is ascribed to M. H. van Aubel [72, 42] as well (Fig. 6.19).

On the sides of a triangle ABC isosceles right triangles ABC', BCA', CAB' are erected. Then $A'B' \perp CC'$ and $|A'B'| = |CC'|$.

The proof immediately follows from the Van Aubel's theorem if we put $C = D$.

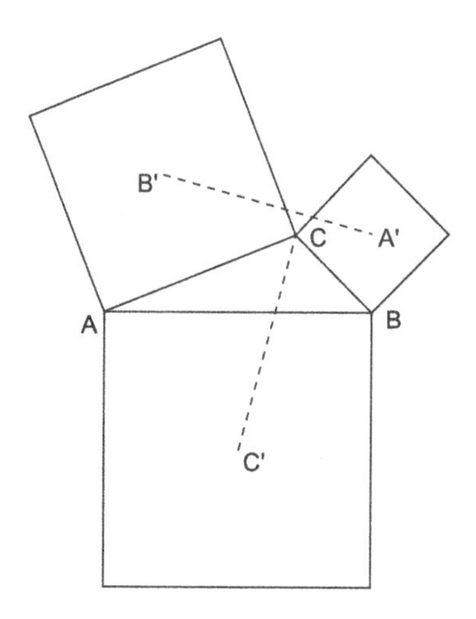

Fig. 6.19 $A'B' \perp CC'$ and $|A'B'| = |CC'|$

Now we will follow the Van Aubel's theorem and consider a special quadrilateral whose diagonals are orthogonal and equal.

Consider a quadrilateral $A'B'C'D'$ whose diagonals are orthogonal and equal, i.e., $A'C' \perp B'D'$, $|A'C'| = |B'D'|$. Denote by A'', B'', C'', D'' the midpoints of sides $A'B', B'C', C'D', D'A'$ respectively. Then $A''B''C''D''$

is a square.

The proof is clear and we will leave it to the reader.

Both of the last theorems can be put together to establish the following theorem:

On the sides of a quadrilateral $ABCD$ construct isosceles right triangles ABA', BCB', CDD', DAD'. Their vertices form a quadrilateral $A'B'C'D'$. Then the midpoints A'', B'', C'', D'' of the sides of $A'B'C'D'$ form a square.

Notice that we need *two operations* to acquire a square from an *arbitrary*

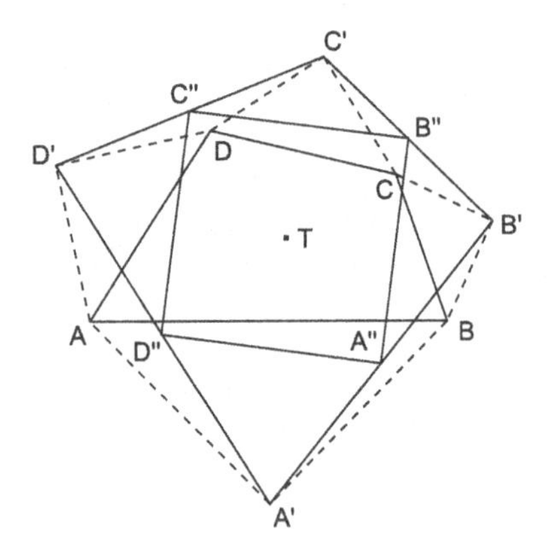

Fig. 6.20 PDN theorem — $A''B''C''D''$ is a square

quadrilateral $ABCD$. In the first step we obtain a (special) quadrilateral $A'B'C'D'$ and in the second step we arrive at a square $A''B''C''D''$ (Fig. 6.20).

Let us show that quadrilaterals $A'B'C'D'$ and $ABCD$ from the PDN theorem have a common centroid T (and $A''B''C''D''$ as well).

Choose a Cartesian coordinate system such that $A = [0, 0]$, $B = [x, 0]$, $C = [y, z]$, $D = [w, t]$ and let $A' = [a_1, a_2]$, $B' = [b_1, b_2]$, $C' = [c_1, c_2]$, $D' = [d_1, d_2]$ (Fig. 6.17). The previous construction of the points A', B', C', D' gives the conditions:

$h_1 : 2a_1 - x = 0,$
$h_2 : 2a_2 + x = 0,$
$h_3 : 2b_1 - x - y - z = 0,$

$h_4 : 2b_2 - z - x + y = 0,$
$h_5 : 2c_1 - y - w - t + z = 0,$
$h_6 : 2c_2 - z - t - y + w = 0,$
$h_7 : 2d_1 - w + t = 0,$
$h_8 : 2d_2 - t - w = 0.$

We will prove that

$\frac{1}{4}(A + B + C + D) = \frac{1}{4}(A' + B' + C' + D') \Leftrightarrow$

$h_9 : x + y + w - a_1 - b_1 - c_1 - d_1 = 0,$
$h_{10} : z + t - a_2 - b_2 - c_2 - d_2 = 0.$

We ask whether the polynomials h_9, h_{10} belong to the ideal $I = (h_1, h_2, \ldots, h_8)$. We enter

```
Use R::=Q[a[1..2]b[1..2]c[1..2]d[1..2]e[1..2]f[1..2]g[1..2]
h[1..2]xyzwt];
I:=Ideal(2a[1]-x,2a[2]+x,2b[1]-x-y-z,2b[2]-z-x+y,2c[1]-y-w-t+z,
2c[2]-z-t-y+w,2d[1]-w+t,2d[2]-t-w);
NF(x+y+w-a[1]-b[1]-c[1]-d[1],I);
```

and get `NF=0`. Similarly we show that $h_{10} \in I$. The theorem is proved.

Another special case of the PDN theorem for a quadrilateral was published by P. Finsler and H. Hadwiger [40]. The Finsler–Hadwiger theorem

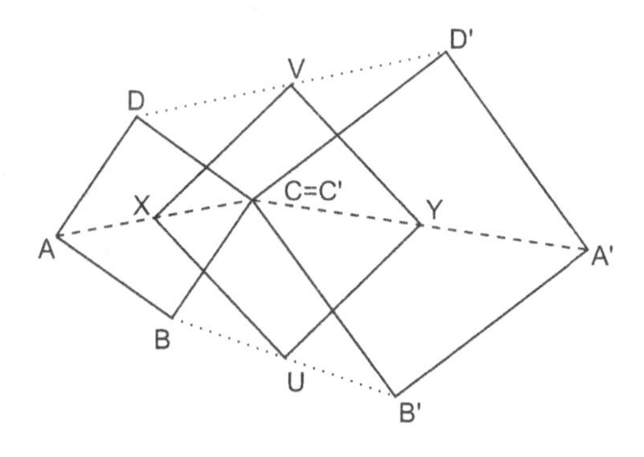

Fig. 6.21 Finsler–Hadwiger theorem: Points X, U, Y, V form a square

reads (Fig. 6.21):

Theorem 6.8 (Finsler–Hadwiger). *Two squares $ABCD$ and $A'B'C'D'$*

with a common vertex $C = C'$ are given. Then the centers X, Y of the squares $ABCD$, $A'B'C'D'$ and the centers U, V of the segments BB', CC' form a square.

The Finsler–Hadwiger theorem follows from the PDN theorem applied to a (degenerate) quadrilateral $ACA'C'$.

In order to obtain a square from an arbitrary quadrilateral in this process of "regularization", we can also exchange the order of both operations:

Given a quadrilateral $ABCD$ in the plane, construct the centers A', B', C', D' of the sides of $ABCD$ which form a parallelogram. On the sides of a parallelogram $A'B'C'D'$ construct isosceles right triangles. Then their vertices A'', B'', C'', D'' form a square.

The proof of this theorem is easy and we will leave it to the reader (the second part contains Thébault's theorem).

Now we can formulate the PDN theorem for quadrilaterals in its general form (Fig. 6.20):

Theorem 6.9 (PDN theorem for $n = 4$). *On the sides of a quadrilateral $ABCD$ construct isosceles triangles with vertex angle $i \cdot 2\pi/4$, where $i \in \{1, 2, 3\}$. Their vertices form a quadrilateral $A'B'C'D'$. On the sides of $A'B'C'D'$ construct isosceles triangles with vertex angle $j \cdot 2\pi/4$, where $j \in \{1, 2, 3\}, j \neq i$. Then the vertices form a regular quadrilateral $A''B''C''D''$. Constructing on the sides of $A''B''C''D''$ isosceles triangles with vertex angle $k \cdot 2\pi/4$, where $k \in \{1, 2, 3\}, k \neq i, j$, we arrive at a point — the common centroid of quadrilaterals $ABCD$, $A'B'C'D'$, $A''B''C''D''$.*

Remark 6.7.

1) Construction of isosceles triangles on the sides of a quadrilateral with the angle $i \cdot 2\pi/4$, where $i \in \{1, 2, 3\}$, admits angles greater then $180°$. For example, by construction of the isosceles triangle ABA' over the side AB of $ABCD$ for the angle $90°$ the vertex A' lies *outside* of a quadrilateral $ABCD$ since the oriented angle of vectors $B - A'$, $A - A'$ equals $90°$, whereas for the value $270°$ the vertex A' is *inside* of $ABCD$, since the oriented angle of vectors $B - A'$, $A - A'$ equals $270°$ (Fig. 6.20).

2) By the PDN theorem for quadrilaterals over the sides of a quadrilateral we construct successively isosceles triangles with vertex angles from the set $\{90°, 180°, 270°\}$ and get various types of theorems according to the order of the angles. However, for *any* choice of the order of the angles this process terminates at one point — the common centroid of all involved

quadrilaterals.

At the end of this section let us give the PDN theorem for n-gons in its general form. It is similar to the just formulated PDN theorem for quadrilaterals. It is as follows [97, 30, 31, 81]:

The PDN theorem for n-gons:

Theorem 6.10 (PDN theorem). *Let $j_1, j_2, \ldots, j_{n-1}$ be any order of numbers $1, 2, \ldots, n-1$. On the sides of an arbitrary plane n-gon $\mathcal{P}$ construct isosceles triangles with vertex angle $j_1 \cdot 2\pi/n$. Resulting vertices form a new polygon $\mathcal{P}_{j_1}$. On the sides of $\mathcal{P}_{j_1}$ construct isosceles triangles with vertex angle $j_2 \cdot 2\pi/n$. We get a polygon $\mathcal{P}_{j_1,j_2}$. Continuing in this construction for all values $j_1, j_2, \ldots, j_{n-1}$, then the final polygon $\mathcal{P}_{j_1,j_2,\ldots,j_{n-1}}$ is a point — the common centroid of all polygons $\mathcal{P}$, $\mathcal{P}_{j_1}$, $\mathcal{P}_{j_1,j_2}$, $\ldots$, $\mathcal{P}_{j_1,j_2,\ldots,j_{n-1}}$, and the n-gon $\mathcal{P}_{j_1,j_2,\ldots,j_{n-2}}$ is j_{n-1}-regular.*

6.3 PDN theorem in space

In this section we will generalize the PDN theorem to n-gons in the three dimensional space. A spatial generalization of the PDN theorem will be shown on a skew pentagon.

6.3.1 *Douglas' pentagon*

In 1960 J. Douglas published a theorem [32] which is a spatial generalization of the PDN theorem. Using this theorem we are able, roughly spoken, to assign to an arbitrary skew pentagon a planar affine-regular pentagon. Douglas' theorem reads:

Theorem 6.11 (Douglas). *Given a skew pentagon $ABCDE$. Denote by M_1, M_2, M_3, M_4, M_5 the midpoints of sides which are opposite to the vertices A, B, C, D, E respectively. On the ray AM_1, outwardly construct a point A_1 such that $|M_1A_1| = 1/\sqrt{5}|AM_1|$. On BM_2, outwardly construct a point B_1 such that $|M_2B_1| = 1/\sqrt{5}|BM_2|$, etc. until we obtain a pentagon $A_1B_1C_1D_1E_1$. Similarly, on the ray AM_1, inwardly construct a point A_1' such that $|M_1A_1'| = 1/\sqrt{5}|AM_1|$, etc., until we obtain a pentagon $A_1'B_1'C_1'D_1'E_1'$. Then the pentagons $A_1B_1C_1D_1E_1$ and $A_1'B_1'C_1'D_1'E_1'$ are planar affine-regular.*

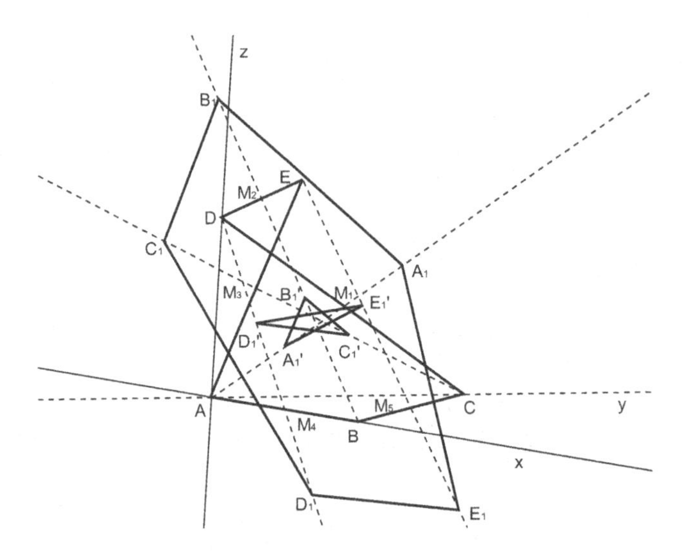

Fig. 6.22 Douglas' pentagon $A_1B_1C_1D_1E_1$

We will "discover" the theorem by computer, i.e. to find a constant k such that the resulting pentagons $A_1B_1C_1D_1E_1$ and $A_1'B_1'C_1'D_1'E_1'$ are planar affine-regular pentagons.

Since we deal with a problem of affine geometry in space (here only ratios of distances occur), we can choose an affine coordinate system such that for the vertices of a pentagon $A = [0,0,0]$, $B = [x,0,0]$, $C = [0,y,0]$, $D = [0,0,z]$, $E = [u,v,w]$ (Fig. 6.22).

Then for the midpoints of the sides $M_1 = [0,y/2,z/2]$, $M_2 = [u/2,v/2,(z+w)/2]$, $M_3 = [u/2,v/2,w/2]$, $M_4 = [x/2,0,0]$, $M_5 = [x/2,y/2,0]$. Further denote $A_1 = [a_1,a_2,a_3]$, $B_1 = [b_1,b_2,b_3]$, $C_1 = [c_1,c_2,c_3]$, $D_1 = [d_1,d_2,d_3]$, $E_1 = [e_1,e_2,e_3]$. On the ray AM_1 outwardly construct a point A_1 such that $A_1 - M_1 = k(M_1 - A)$, where k is an unknown real number for the moment. Similarly we acquire the other points B_1, C_1, D_1, E_1. We get the following relations:

$A_1 - M_1 = k(M_1 - A) \Leftrightarrow$
$h_1 : a_1 = 0,\ h_2 : 2a_2 - y - ky = 0,\ h_3 : 2a_3 - z - kz = 0,$

$B_1 - M_2 = k(M_2 - B) \Leftrightarrow$
$h_4 : 2b_1 - u - k(u - 2x) = 0, h_5 : 2b_2 - v - kv = 0, h_6 : 2b_3 - z - w - k(z + w) = 0,$

$C_1 - M_3 = k(M_3 - C) \Leftrightarrow$

$h_7: 2c_1 - u - ku = 0, h_8: 2c_2 - v - k(v-2y) = 0, h_9: 2c_3 - w - kw = 0,$

$D_1 - M_4 = k(M_4 - D) \Leftrightarrow$
$h_{10}: 2d_1 - x - kx = 0,\ h_{11}: d_2 = 0,\ h_{12}: d_3 + kz = 0,$

$E_1 - M_5 = k(M_5 - E) \Leftrightarrow$
$h_{13}: 2e_1 - x - k(x-2u) = 0, h_{14}: 2e_2 - y - k(y-2v) = 0, h_{15}: e_3 + kw = 0.$

We are looking for a real k such that the points A_1, B_1, C_1, D_1, E_1 form an affine-regular pentagon. We know that a pentagon $A_1B_1C_1D_1E_1$ is affine-regular iff

$$A_1B_1 \parallel C_1E_1,\ B_1C_1 \parallel D_1A_1, C_1D_1 \parallel E_1B_1, D_1E_1 \parallel A_1C_1, E_1A_1 \parallel B_1D_1 \tag{6.14}$$

(see [129]). First we will suppose that $A_1B_1 \parallel C_1E_1$ which means that:

$h_{16}: (b_1 - a_1)(c_2 - e_2) - (b_2 - a_2)(c_1 - e_1) = 0,$
$h_{17}: (b_1 - a_1)(c_3 - e_3) - (b_3 - a_3)(c_1 - e_1) = 0,$
$h_{18}: (b_2 - a_2)(c_3 - e_3) - (b_3 - a_3)(c_2 - e_2) = 0.$

In the ideal $I = (h_1, h_2, \ldots, h_{18})$ we eliminate all variables except x, y, z, u, v, k. We get

```
Use R::=Q[xyzuvwa[1..3]b[1..3]c[1..3]d[1..3]e[1..3]k];
I:=Ideal(a[1],2a[2]-y-ky,2a[3]-z-kz,2b[1]-u-k(u-2x),2b[2]-v-kv,
2b[3]-z-w-k(z+w),2c[1]-u-ku,2c[2]-v-k(v-2y),2c[3]-w-kw,2d[1]-x
-kx,d[2],d[3]+kz,2e[1]-x-k(x-2u),2e[2]-y-k(y-2v),e[3]+kw,(b[1]-
a[1])(e[2]-c[2])-(b[2]-a[2])(e[1]-c[1]),(b[1]-a[1])(e[3]-c[3])-
(b[3]-a[3])(e[1]-c[1]),(b[2]-a[2])(e[3]-c[3])-(b[3]-a[3])(e[2]-
c[2]));
Elim(w..e[3],I);
```

the condition

$$(5k^2 - 1)(y - v)x = 0.$$

In an analogous way, for $B_1C_1 \parallel D_1A_1$ we get the condition

$$(5k^2 - 1)xz = 0,$$

etc. We obtain two values $k = 1/\sqrt{5}, k = -1/\sqrt{5}$ which come into consideration.

Now suppose that the values $k = 1/\sqrt{5},\ k = -1/\sqrt{5}$ are given. We are to show that then the conditions (6.14) are fulfilled. We show that for example $D_1E_1 \parallel A_1C_1$. We enter

```
Use R::=Q[xyzuvwa[1..3]b[1..3]c[1..3]d[1..3]e[1..3]k];
J:=Ideal(a[1],2a[2]-y-ky,2a[3]-z-kz,2b[1]-u-k(u-2x),2b[2]-v-kv,
2b[3]-z-w-k(z+w),2c[1]-u-ku,2c[2]-v-k(v-2y),2c[3]-w-kw,2d[1]-x
-kx,d[2],d[3]+kz,2e[1]-x-k(x-2u),2e[2]-y-k(y-2v),e[3]+kw,5k^2-1);
NF((e[1]-d[1])(c[2]-a[2])-(e[2]-d[2])(c[1]-a[1]),J);
```

and get the answer `NF=0`. Continuing in this verification we conclude that for both values $k = \pm 1/\sqrt{5}$ we get affine-regular pentagons.

For $k = 1/\sqrt{5}$ we get a *convex* affine-regular pentagon $A_1B_1C_1D_1E_1$, whereas the value $k = -1/\sqrt{5}$ gives a *non-convex* star affine-regular pentagon $A_1'B_1'C_1'D_1'E_1'$ (Fig. 6.22).

Remark 6.8.
1) Both affine-regular pentagons $A_1B_1C_1D_1E_1$ and $A_1'B_1'C_1'D_1'E_1'$ lie in different planes which pass through the common centroid of both affine-regular pentagons and the original pentagon $ABCDE$.

2) Douglas' theorem for a skew pentagon can be generalized to an arbitrary skew n-gon [117, 88, 91].

Chapter 7

Geometric inequalities

In this chapter we will study some geometric inequalities which can be discovered and proved by computer using Gröbner bases computation. We will use the method which is based on the decomposition of a polynomial into the sum of squares.

Adopting an appropriate coordinate system, we eliminate dependent variables, and transform a given expression into such a form from which its non-negativity (or non-positivity) follows. Most often a given formula is expressed in the form of a sum of squares of polynomials of independent variables from which non-negativity follows. The sum of squares (sos) decomposition of a polynomial is closely connected with the 17th Hilbert's problem. See [106] by B. Reznick, where the survey of this problem and possible techniques to solve it are given. See also [21, 86].

First we will be concerned with the so-called parallelogram law. Then we will generalize the parallelogram law on the inequality between the sides and diagonals of a polygon. Next we will be concerned with the Neuberg–Pedoe inequality which is related to two triangles with given side lengths and areas. Then the well-known inequality between the sum of squares of sides and the area of a polygon and the discrete case of Wirtinger's inequality of polygons are investigated. The isoperimetric inequality for a triangle concludes the part, where the same technique of automatic proving is used.

At the end of this chapter we will show another approach to solve geometric inequalities without the use of a coordinate system. This is based on the expression of a given inequality in terms of other geometric magnitudes from which the inequality is seen. Using this method the inequality of Euler is solved.

Besides computer proofs we shall show classical solutions as well, to compare both approaches.

7.1 Inequality between the diagonals of an n-gon

The best known geometric inequality is probably the triangle inequality which says that the sum of two sides of a triangle is greater than the third side. We will deal with the inequality between the sides and diagonals of an n-gon. To simplify the situation we will use the word "diagonal" for any segment joining two vertices of an n-gon. First look at a quadrilateral. We start with an investigation of equalities between diagonals holding for various types of quadrilaterals.

7.1.1 *Parallelogram law*

In this part we will explore the equality between the sum of squares of sides and the sum of squares of diagonals of a parallelogram. This equality is known as the *parallelogram law* [44]:

Theorem 7.1 (Parallelogram law). *Given a parallelogram with the side lengths a, b and diagonals e, f. Then*

$$2(a^2 + b^2) = e^2 + f^2. \tag{7.1}$$

Let us prove the relation (7.1). Denote the vertices of a parallelogram by the letters A, B, C, D and its side lengths and diagonals by $a = |AB| = |CD|$, $b = |BC| = |DA|$, $e = |AC|$, $f = |BD|$. Choose a Cartesian coordinate

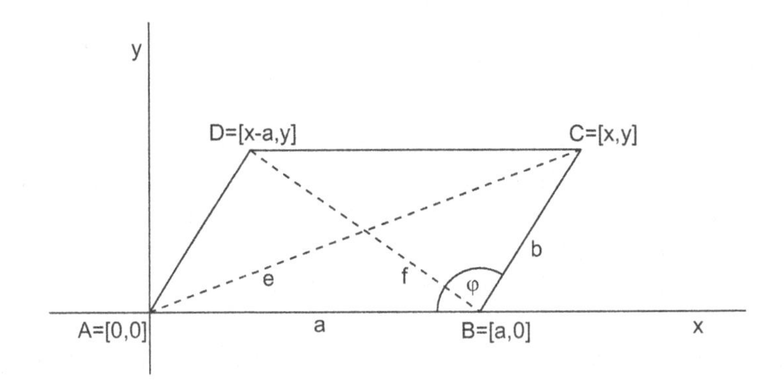

Fig. 7.1 Parallelogram law: $2(a^2 + b^2) = e^2 + f^2$

system such that $A = [0,0]$, $B = [a,0]$, $C = [x,y]$, $D = [x-a,y]$ (Fig. 7.1). Then

$b = |BC| = |AD| \Leftrightarrow h_1 : (x-a)^2 + y^2 - b^2 = 0,$
$e = |AC| \Leftrightarrow h_2 : x^2 + y^2 - e^2 = 0,$
$f = |BD| \Leftrightarrow h_3 : (x-2a)^2 + y^2 - f^2 = 0.$

The conclusion c of a statement has the form

$c : 2(a^2 + b^2) - (e^2 + f^2) = 0.$

We compute the normal form of c with respect to the Gröbner basis of the ideal $I = (h_1, h_2, h_3)$ for some prescribed order of variables. In CoCoA we enter

```
Use R::=Q[axybeft];
I:=Ideal((x-a)^2+y^2-b^2,x^2+y^2-e^2,(x-2a)^2+y^2-f^2,
(2(a^2+b^2)-(e^2+f^2))t-1);
NF(1,I);
```

and get zero which means that from the assumptions h_1, h_2, h_3 the conclusion c follows. Theorem 7.1 is hence proved.

A classical proof of the equality (7.1) is as follows:
By the law of cosines in the triangle ABC it holds (Fig. 7.1),

$$e^2 = a^2 + b^2 - 2ab\cos\varphi.$$

Analogously in the triangle ABD we have

$$f^2 = a^2 + b^2 - 2ab\cos(\pi - \varphi).$$

When adding both equalities we get (7.1).

Remark 7.1. The parallelogram law (7.1) was already known in a slightly different form to ancient Greeks. It became familiar since the year 1935, when J. von Neumann showed that the Banach space in which (7.1) holds is the Hilbert space.

The relation (7.1) is a necessary condition for a quadrilateral $ABCD$ to be a parallelogram. Is the condition (7.1) also a sufficient condition? We will prove the following theorem [44]:

Theorem 7.2. *A quadrilateral $ABCD$ with side lengths a, b, c, d and diagonals e, f is given. Then $ABCD$ is a parallelogram if and only if*

$$a^2 + b^2 + c^2 + d^2 = e^2 + f^2. \tag{7.2}$$

The " only if " part of this theorem has already been proved. Now assume that (7.2) holds. We will explore all quadrilaterals which fulfil (7.2). We are to show that $ABCD$ is a parallelogram.

Let us denote $a = |AB|$, $b = |BC|$, $c = |CD|$, $d = |DA|$, $e = |AC|$, $f = |BD|$ and choose a Cartesian system of coordinates so that $A = [0, 0]$, $B = [a, 0]$, $C = [x, y]$, $D = [u, v]$ (Fig. 7.2). Algebraic translation of the

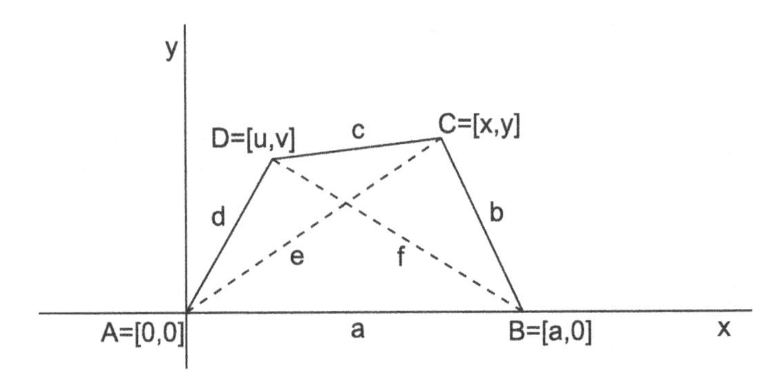

Fig. 7.2 $a^2 + b^2 + c^2 + d^2 = e^2 + f^2 \Leftrightarrow ABCD$ is a parallelogram

assumptions is as follows:

$|BC| = b \Leftrightarrow h_1 : (x - a)^2 + y^2 - b^2 = 0,$
$|CD| = c \Leftrightarrow h_2 : (u - x)^2 + (v - y)^2 - c^2 = 0,$
$|DA| = d \Leftrightarrow h_3 : u^2 + v^2 - d^2 = 0,$
$|AC| = e \Leftrightarrow h_4 : x^2 + y^2 - e^2 = 0,$
$|BD| = f \Leftrightarrow h_5 : (u - a)^2 + v^2 - f^2 = 0,$
$h_6 : a^2 + b^2 + c^2 + d^2 - e^2 - f^2 = 0.$

In the ideal $I = (h_1, h_2, \ldots, h_6)$ we eliminate *dependent* variables b, c, d, e, f. We get

```
Use R::=Q[xyuvabcdef];
I:=Ideal((x-a)^2+y^2-b^2,(u-x)^2+(v-y)^2-c^2,u^2+v^2-d^2,x^2
+y^2-e^2,(u-a)^2+v^2-f^2,a^2+b^2+c^2+d^2-e^2-f^2);
Elim(b..f,I);
```

the condition $x^2 + y^2 - 2xu + u^2 - 2yv + v^2 - 2xa + 2ua + a^2 = 0$ which can be expressed in the form

$$(x - u - a)^2 + (y - v)^2 = 0. \tag{7.3}$$

The condition (7.3) means that a quadrilateral is a parallelogram since for real numbers x, y, u, v, a the condition (7.3) implies $x - u - a = 0$ and $y - v = 0$, i.e. $A - D = B - C$. A verification

```
Use R::=Q[xyuvabcdef];
I:=Ideal((x-a)^2+y^2-b^2,(u-x)^2+(v-y)^2-c^2,u^2+v^2-d^2,x^2
+y^2-e^2,(u-a)^2+v^2-f^2,a^2+b^2+c^2+d^2-e^2-f^2);
NF((x-u-a)^2+(y-v)^2,I);
0
```

confirms that the normal form equals zero. Hence a quadrilateral which fulfils $a^2+b^2+c^2+d^2=e^2+f^2$ is a parallelogram. Theorem 7.2 is proved.

Remark 7.2. If we enter $x-u-a=y-v=0$ instead of the "equivalent" condition $(x-u-a)^2+(y-v)^2=0$, which also characterizes a parallelogram, we fail. What is the difference between the conditions

$$(x-u-a)^2+(y-v)^2=0 \quad \text{and} \quad x-u-a=0 \ \wedge \ y-v=0\ ?$$

"The only difference" is the fact that whereas in the real case both conditions are the same, in complex numbers

$$(x-u-a)^2+(y-v)^2=(x-u-a+i(y-v))(x-u-a-i(y-v)),$$

and we get two conditions

$$x-u-a+i(y-v)=0, \quad x-u-a-i(y-v)=0.$$

This example should be instructive. We always have to consider such conditions which the system requires (and yields). Any "change" of these conditions, despite the fact that they describe the reality properly, mostly leads to the failure.

The parallelogram law (7.1) can be generalized on a trapezoid [82]:

Theorem 7.3. *A quadrilateral $ABCD$ is a trapezoid with bases a, c, legs b, d and diagonals e, f if and only if*

$$b^2+d^2+2ac=e^2+f^2. \tag{7.4}$$

A proof by computer is as follows. First prove that the condition (7.4) is necessary. Let us suppose that $ABCD$ is a trapezoid.

Let $a=|AB|$, $b=|BC|$, $c=|CD|$, $d=|DA|$, $e=|AC|$, $f=|BD|$. Choose a Cartesian coordinate system so that $A=[0,0]$, $B=[a,0]$, $C=[x,y]$, $D=[u,v]$ (Fig. 7.3). Then

$b=|BC| \Leftrightarrow h_1:(x-a)^2+y^2-b^2=0,$
$c=|CD| \Leftrightarrow h_2:(u-x)^2+(v-y)^2-c^2=0,$
$d=|DA| \Leftrightarrow h_3:u^2+v^2-d^2=0,$
$e=|AC| \Leftrightarrow h_4:x^2+y^2-e^2=0,$

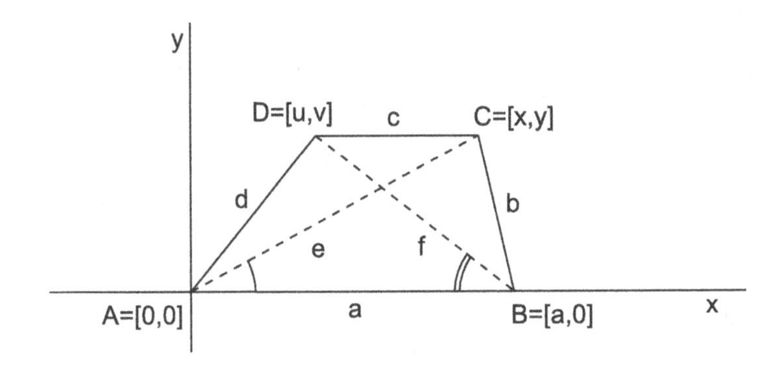

Fig. 7.3 In a trapezoid $b^2 + d^2 + 2ac = e^2 + f^2$ holds

$f = |BD| \Leftrightarrow h_5 : (u-a)^2 + v^2 - f^2 = 0,$
$AB \parallel CD \Leftrightarrow h_6 : y - v = 0.$

The conclusion z has the form

$z : b^2 + d^2 + 2ac - e^2 - f^2 = 0.$

Computing the normal form of 1 in the ideal $I = (h_1, h_2, \ldots, h_6, zt-1)$ we obtain

```
Use R::=Q[abcdefxyuvt];
I:=Ideal((x-a)^2+y^2-b^2,(u-x)^2+(v-y)^2-c^2,u^2+v^2-d^2,x^2+
y^2-e^2,(u-a)^2+v^2-f^2,y-v,(b^2+d^2+2ac-e^2-f^2)t-1);
NF(1,I);
```

the answer 1. Hence the statement (7.4) is not generally true. Therefore in the ideal I we eliminate a slack variable t to find out additional conditions. We get

```
Use R::=Q[abcdefxyuvt];
I:=Ideal((x-a)^2+y^2-b^2,(u-x)^2+(v-y)^2-c^2,u^2+v^2-d^2,x^2+
y^2-e^2,(u-a)^2+v^2-f^2,y-v,(b^2+d^2+2ac-e^2-f^2)t-1);
Elim(t,I);
```

the condition $x - u + c = 0$ which must be ruled out. Namely, if the condition $x - u + c = 0$ holds, then $ABCD$ intersects itself. We get

```
Use R::=Q[abcdefxyuvst];
J:=Ideal((x-a)^2+y^2-b^2,(u-x)^2+(v-y)^2-c^2,u^2+v^2-d^2,x^2+
y^2-e^2,(u-a)^2+v^2-f^2,y-v,(x-u+c)s-1,(b^2+d^2+2ac-e^2-f^2)t
```

```
-1);
NF(1,J);
```

the answer 0 and (7.4) is true.

Now we will show that (7.4) is sufficient for a quadrilateral $ABCD$ to be a trapezoid. First, similarly as in the previous part, we find additional conditions which are of the form $a(x-u+c)s-1=0$. The geometric meaning is that $A \neq B$ and $ABCD$ does not intersect itself. Adding these conditions to the respective ideal we get

```
Use R::=Q[abcdefxyuvst];
K:=Ideal((x-a)^2+y^2-b^2,(u-x)^2+(v-y)^2-c^2,u^2+v^2-d^2,x^2+
y^2-e^2,(u-a)^2+v^2-f^2,b^2+d^2+2ac-e^2-f^2,a(x-u+c)s-1,
(y-v)t-1);
NF(1,K);
```

the result 0 which implies that $ABCD$ is a trapezoid.

A classical proof of (7.4) is as follows (Fig. 7.3).
In the triangle ABC by the law of cosines

$$b^2 = a^2 + e^2 - 2ae\cos(\angle CAB)$$

holds. Similarly, in the triangle ABD

$$d^2 = a^2 + f^2 - 2af\cos(\angle ABD).$$

When adding the left and right sides we acquire

$$b^2 + d^2 = e^2 + f^2 + 2a^2 - 2a(e\cos(\angle CAB) + f\cos(\angle ABD)),$$

which, with respect to the equality $e\cos(\angle CAB) + f\cos(\angle ABD) = a + c$, gives (7.4).

7.1.2 *Case of a quadrilateral*

In the previous part we showed that the equality (7.2) holds only for a parallelogram. Now we will generalize Theorem 7.3 [44]:

Theorem 7.4. *Given a quadrilateral $ABCD$ with side lengths a, b, c, d and diagonals e, f. Then*

$$a^2 + b^2 + c^2 + d^2 \geq e^2 + f^2. \tag{7.5}$$

The sign of equality in (7.5) is attained if and only if $ABCD$ is a parallelogram.

We see that a special case of (7.5) is just the parallelogram law (7.1). The inequality between the sides and diagonals of a quadrilateral (7.5) holds even for a skew quadrilateral $ABCD$ as we will show later.

Let us denote $a = |AB|, b = |BC|, c = |CD|, d = |DA|, e = |AC|, f = |BD|$. Choose a Cartesian system of coordinates so that $A = [0,0]$, $B = [a,0]$, $C = [x,y]$, $D = [u,v]$ (Fig. 7.4). Then

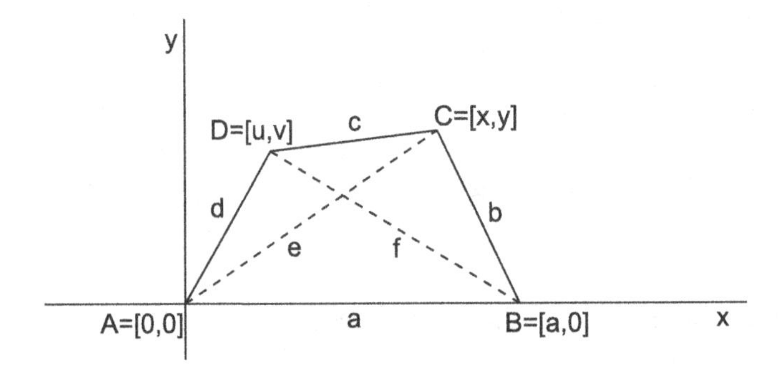

Fig. 7.4 In a quadrilateral $ABCD$: $a^2 + b^2 + c^2 + d^2 \geq e^2 + f^2$

$$|BC| = b \Leftrightarrow h_1 : (x-a)^2 + y^2 - b^2 = 0,$$
$$|CD| = c \Leftrightarrow h_2 : (u-x)^2 + (v-y)^2 - c^2 = 0,$$
$$|DA| = d \Leftrightarrow h_3 : u^2 + v^2 - d^2 = 0,$$
$$|AC| = e \Leftrightarrow h_4 : x^2 + y^2 - e^2 = 0,$$
$$|BD| = f \Leftrightarrow h_5 : (u-a)^2 + v^2 - f^2 = 0.$$

To prove (7.5) we express the difference $(a^2 + b^2 + c^2 + d^2) - (e^2 + f^2)$ as a sum of squares of polynomials of *independent* variables from which its non-negativity follows.

For this reason we introduce a slack variable k such that

$$h_6 : (a^2 + b^2 + c^2 + d^2) - (e^2 + f^2) - k = 0. \tag{7.6}$$

We eliminate dependent variables b, c, d, e, f in the ideal $I = (h_1, h_2, \ldots, h_6)$. CoCoA returns

```
Use R::=Q[xyuvabcdefk];
I:=Ideal((x-a)^2+y^2-b^2,(u-x)^2+(v-y)^2-c^2,u^2+v^2-d^2,x^2+
y^2-e^2,(u-a)^2+v^2-f^2,a^2+b^2+c^2+d^2-e^2-f^2-k);
Elim(b..f,I);
```

the polynomial $x^2 + y^2 - 2xu + u^2 - 2yv + v^2 - 2xa + 2ua + a^2 - k$ which can be expressed in the form

$$k = (x - u - a)^2 + (y - v)^2. \tag{7.7}$$

Substitution of k from (7.7) into (7.6) gives the following identity

$$a^2 + b^2 + c^2 + d^2 - e^2 - f^2 = (x - u - a)^2 + (y - v)^2, \tag{7.8}$$

which implies the inequality (7.5).

The equality in (7.5) is attained if and only if $x-u-a = 0$ and $y-v = 0$, that is, $ABCD$ is a parallelogram as we could see in the previous part.

The inequality (7.5) also follows from the next theorem which is ascribed to L. Euler [66], Fig. 7.5, and which can be considered as a classical proof of Theorem 7.4.

Theorem 7.5. *Given a quadrilateral $ABCD$. Denote P, Q as the mid-points of the diagonals AC and BD respectively. Then*

$$|AB|^2 + |BC|^2 + |CD|^2 + |DA|^2 = |AC|^2 + |BD|^2 + 4|PQ|^2. \tag{7.9}$$

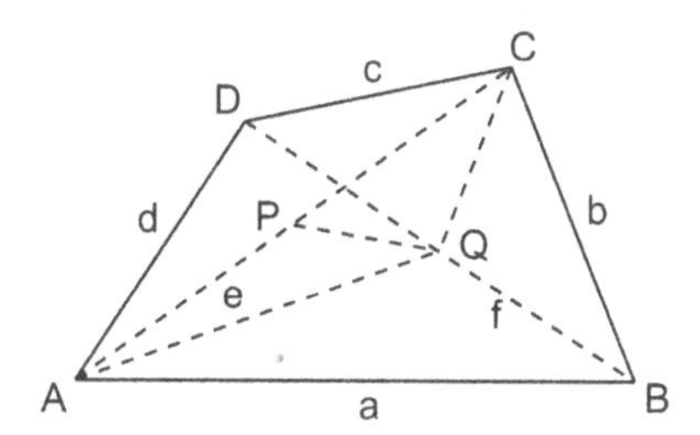

Fig. 7.5 In a quadrilateral $ABCD$: $|AB|^2 + |BC|^2 + |CD|^2 + |DA|^2 = |AC|^2 + |BD|^2 + 4|PQ|^2$

By Stewart's theorem [4] we express the length of a median AQ of a triangle ABD knowing its side lengths

$$4|AQ|^2 = 2|AB|^2 + 2|AD|^2 - |BD|^2.$$

Analogously in the triangle BCD we get

$$4|CQ|^2 = 2|BC|^2 + 2|CD|^2 - |BD|^2.$$

When adding both relations we obtain

$$|AB|^2 + |BC|^2 + |CD|^2 + |DA|^2 = |BD|^2 + 2|AQ|^2 + 2|CQ|^2.$$

In the triangle ACQ by the same theorem

$$4|PQ|^2 = 2|AQ|^2 + 2|CQ|^2 - |AC|^2.$$

Substitution into the previous relation gives (7.9).

From (7.9) the inequality (7.5) follows. The sign of equality in (7.5) is attained if and only if $|PQ| = 0$, i.e. for a parallelogram. The theorem is proved.

Remark 7.3.

1) The relation (7.9) and thus also the relation (7.5) are valid for an arbitrary *skew* quadrilateral as we can see from the previous proof. Namely, a quadrilateral $ABCD$ in Fig. 7.5 can be viewed as a skew quadrilateral. By this all steps of the proof were preserved.

2) Note that for the expression on the right-hand side of (7.8)

$$(x - u - a)^2 + (y - v)^2 = |(C - B) - (D - A)|^2 = 4|P - Q|^2$$

holds, since (Fig. 7.4)

$$P = (A + C)/2, \ Q = (B + D)/2 \text{ and } |P - Q| = 1/2|(A + C) - (B + D)|.$$

3) Let $ABCD$ be a skew quadrilateral. The relation (7.9) can also be

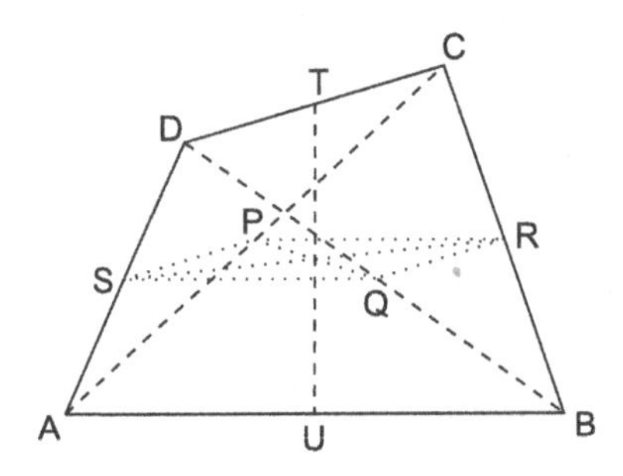

Fig. 7.6 In a skew quadrilateral $ABCD$ the equality $|AB|^2 + |BC|^2 + |CD|^2 + |DA|^2 = |AC|^2 + |BD|^2 + 4|PQ|^2$ holds

proved in the following way (Fig. 7.6).

Let $a = |AB|$, $b = |BC|$, $c = |CD|$, $d = |DA|$, $e = |AC|$, $f = |BD|$. Denote by P, Q the midpoints of the diagonals AC, BD and let R, T, S, U be the midpoints of the sides BC, CD, DA, AB respectively. Then $QRPS$ is a parallelogram in which

$$|QR| = |SP| = \frac{c}{2} \text{ and } |RP| = |SQ| = \frac{a}{2}.$$

By (7.1), $2(|SQ|^2 + |QR|^2) = |PQ|^2 + |SR|^2$ implies

$$\frac{a^2}{2} + \frac{c^2}{2} = |PQ|^2 + |SR|^2. \tag{7.10}$$

Analogously from parallelograms $PUQT$ and $RTSU$ we get

$$\frac{b^2}{2} + \frac{d^2}{2} = |PQ|^2 + |TU|^2 \tag{7.11}$$

and

$$\frac{e^2}{2} + \frac{f^2}{2} = |SR|^2 + |TU|^2. \tag{7.12}$$

The sum of the first two relations (7.10), (7.11) with the following subtraction of the third equality (7.12) give (7.9).

7.1.3 *General case*

Theorem 7.4 is a special case of the following theorem which was published in 1980 by L. Gerber [44]:

Theorem 7.6 (Gerber). *Let $\mathcal{P} = P_0P_1 \ldots P_{n-1}$ be a closed skew n-gon in the Euclidean space E^d. Then*

$$\sum_{k=0}^{n-1} |P_kP_{k+2}|^2 \;\leq\; 4\cos^2\frac{\pi}{n}\sum_{k=0}^{n-1}|P_kP_{k+1}|^2, \tag{7.13}$$

with the equality if and only if $\mathcal{P}$ is a planar affine-regular n-gon.

For $n = 4$ we get the relation (7.5) since an affine image of a square is a parallelogram. Another generalization of the inequality (7.13) was published in 1990 (see [87]). It is as follows:

Theorem 7.7. *Let $\mathcal{P} = P_0P_1 \ldots P_{n-1}$ be a closed skew n-gon in the Euclidean space E^d and let $P_{n+j} = P_j$, for $j = 0, 1, \ldots$. Then for all $p = 1, \ldots, n-1$*

$$\sin^2\frac{\pi}{n}\sum_{k=0}^{n-1}|P_kP_{k+p}|^2 \;\leq\; \sin^2(p\frac{\pi}{n})\sum_{k=0}^{n-1}|P_kP_{k+1}|^2. \tag{7.14}$$

The equality for $1 < p < n-1$ is attained if and only if $\mathcal{P}$ is a planar affine-regular n-gon.

For $p = 2$ we get from (7.14) the inequality (7.13).

7.2 Neuberg–Pedoe inequality

In this section we will prove the well-known Neuberg–Pedoe inequality [95, 75]. The theorem reads:

Theorem 7.8 (Neuberg–Pedoe). *Let a, b, c and k, l, m be the side lengths of the triangles ABC and KLM respectively. Denote by p the area of ABC and by q the area of KLM. Then*

$$k^2(-a^2+b^2+c^2)+l^2(a^2-b^2+c^2)+m^2(a^2+b^2-c^2) \geq 16pq. \quad (7.15)$$

The equality in (7.15) is attained if and only if the triangles ABC and KLM are similar.

Since the triangles ABC, KLM can be situated arbitrarily, we place them into a Cartesian coordinate system so that $A = [x, y]$, $B = [0, 0]$, $C = [a, 0]$, $K = [u, v]$, $L = [0, 0]$, $M = [k, 0]$ (Fig. 7.7). We express the side lengths $a = |BC|$, $b = |CA|$, $c = |AB|$, $k = |LM|$, $l = |MK|$, $m = |KL|$ and the areas p, q in the form of algebraic equations:

$b = |CA| \Leftrightarrow h_1 : (x-a)^2 + y^2 - b^2 = 0,$

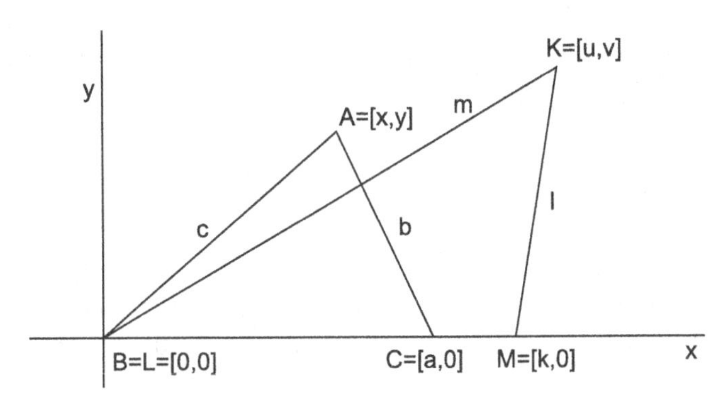

Fig. 7.7 Neuberg–Pedoe inequality

$c = |AB| \Leftrightarrow h_2 : x^2 + y^2 - c^2 = 0,$

$l = |MK| \Leftrightarrow h_3 : (u-k)^2 + v^2 - l^2 = 0,$

$m = |KL| \Leftrightarrow h_4 : u^2 + v^2 - m^2 = 0,$

$p =$ area of $ABC \Leftrightarrow h_5 : 2p - ay = 0,$

$q =$ area of $KLM \Leftrightarrow h_6 : 2q - kv = 0.$

Let t be a slack variable which satisfies

$$h_7 : k^2(-a^2+b^2+c^2)+l^2(a^2-b^2+c^2)+m^2(a^2+b^2-c^2)-16pq-t=0.$$

We want to express t in terms of *independent* variables x, y, a, u, v, k in such a form from which its non-negativity follows. In the ideal $I = (h_1, h_2, \ldots, h_7)$ we eliminate *dependent* variables b, c, l, m, p, q to obtain

```
Use R::=Q[xyuvakbclmpqt];
I:=Ideal((x-a)^2+y^2-b^2,x^2+y^2-c^2,(u-k)^2+v^2-l^2,
u^2+v^2-m^2,2p-ay,2q-kv,
k^2(-a^2+b^2+c^2)+l^2(a^2-b^2+c^2)+m^2(a^2+b^2-c^2)-16pq-t);
Elim(b..q,I);
```

a polynomial which leads to the equation

$$2u^2a^2+2v^2a^2-4xuak-4yvak+2x^2k^2+2y^2k^2-t=0. \tag{7.16}$$

From (7.16) we get

$$t=2(xk-ua)^2+2(yk-va)^2. \tag{7.17}$$

We see that t is in the form of a sum of squares from which the validity of (7.15) follows. The equality in (7.15) is attained iff $xk-ua=0$ and $yk-va=0$, or equivalently $a/k=x/u=y/v$, which means that the triangles ABC and KLM are similar. The theorem is proved.

For a classical proof of Theorem 7.8 see the classical proof of Theorem 6.3 from Chapter 6 (relation (6.12)).

Remark 7.4. If KLM is equilateral then the inequality (7.15) becomes the sum of squares inequality (6.8) for a triangle

$$a^2+b^2+c^2-4\sqrt{3}\,p \geq 0$$

which we encountered in Chapter 6 (see also Section 7.4).

7.3 Wirtinger's inequality

Another inequality — the Wirtinger's inequality [8, 54] — is mostly known from analysis. The discrete case of the Wirtinger's inequality for polygons reads [37, 115]:

Theorem 7.9. *Let $\mathcal{P} = A_0A_1\ldots A_{n-1}$ be a polygon in a plane, where all subscripts are modulo n. Let $A_0 + A_1 + \cdots + A_{n-1} = 0$. Then*

$$\sum_{k=0}^{n-1} |A_kA_{k+1}|^2 \geq 4\sin^2\frac{\pi}{n}\sum_{k=0}^{n-1}|A_k|^2. \tag{7.18}$$

The equality is attained iff $\mathcal{P}$ is an affine-regular n-gon.

We will prove the theorem for $n = 4$ by computer.

Let $A_0A_1A_2A_3$ be an arbitrary quadrilateral. Choose a Cartesian coordinate system so that $A_0 = [x_1, x_2]$, $A_1 = [y_1, y_2]$, $A_2 = [u_1, u_2]$, $A_3 = [v_1, v_2]$ (Fig. 7.8). Denoting $a = |A_0A_1|$, $b = |A_1A_2|$, $c = |A_2A_3|$, $d = |A_3A_0|$, we get the following set of algebraic equations:

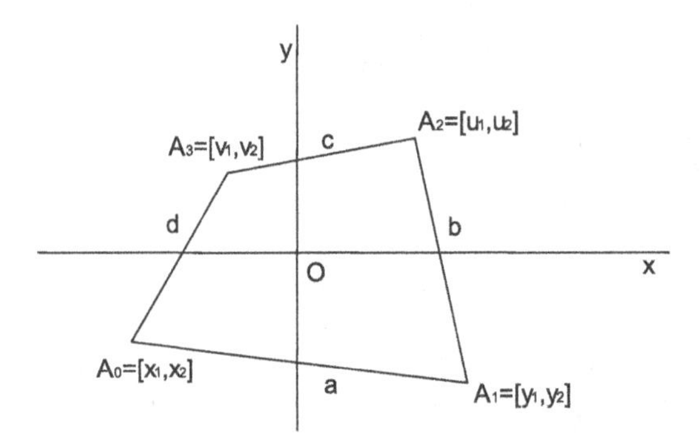

Fig. 7.8 Wirtinger's inequality

$a = |A_0A_1| = |A_0 - A_1| \Leftrightarrow h_1 : (x_1 - y_1)^2 + (x_2 - y_2)^2 - a^2 = 0,$

$b = |A_1A_2| = |A_1 - A_2| \Leftrightarrow h_2 : (y_1 - u_1)^2 + (y_2 - u_2)^2 - b^2 = 0,$

$c = |A_2A_3| = |A_2 - A_3| \Leftrightarrow h_3 : (u_1 - v_1)^2 + (u_2 - v_2)^2 - c^2 = 0,$

$d = |A_3A_0| = |A_3 - A_0| \Leftrightarrow h_4 : (v_1 - x_1)^2 + (v_2 - x_2)^2 - d^2 = 0,$

$A_0 + A_1 + A_2 + A_3 = 0 \Leftrightarrow$

$h_5 : x_1 + y_1 + u_1 + v_1 = 0,\ h_6 : x_2 + y_2 + u_2 + v_2 = 0.$

Now we introduce a slack variable t such that

$\sum_{k=0}^{3} |A_k - A_{k+1}|^2 - 4\sin^2\frac{\pi}{4}\sum_{k=0}^{3}|A_k|^2 - t = 0 \Leftrightarrow$

$h_7 : a^2 + b^2 + c^2 + d^2 - 2(x_1^2 + x_2^2 + y_1^2 + y_2^2 + u_1^2 + u_2^2 + v_1^2 + v_2^2) - t = 0.$

Elimination of v_1, v_2, a, b, c, d in the ideal $I = (h_1, h_2, \ldots, h_7)$ gives

```
Use R::=Q[x[1..2]y[1..2]u[1..2]v[1..2]abcdt];
I:=Ideal((x[1]-y[1])^2+(x[2]-y[2])^2-a^2,(y[1]-u[1])^2+(y[2]-
u[2])^2-b^2,(u[1]-v[1])^2+(u[2]-v[2])^2-c^2,(v[1]-x[1])^2+
(v[2]-x[2])^2-d^2,x[1]+y[1]+u[1]+v[1],x[2]+y[2]+u[2]+v[2],a^2
+b^2+c^2+d^2-2(x[1]^2+x[2]^2+y[1]^2+y[2]^2+u[1]^2+u[2]^2+
v[1]^2+v[2]^2)-t);
Elim(v[1]..d,I);
```

the polynomial equation

$$-2x_1^2 - 2x_2^2 - 4x_1u_1 - 2u_1^2 - 4x_2u_2 - 2u_2^2 + t = 0,$$

in which t can be expressed in the form

$$t = 2(x_1 + u_1)^2 + 2(x_2 + u_2)^2. \tag{7.19}$$

Thus the inequality (7.18) for $n = 4$ is proved. The equality is attained iff $x_1 + u_1 = x_2 + u_2 = 0$, that is, $x_1 = -u_1, x_2 = -u_2$. Then from $h_5 = 0$ and $h_6 = 0$ relations $y_1 = -v_1, y_2 = -v_2$ follow and $A_1A_2A_3A_4$ is a parallelogram.

To prove Wirtinger's inequality classically we will use the following vector method. Place the centroid O of $A_1A_1A_2A_3$ into the origin of a respective vector space. Let $U \cdot V$ be a scalar product of U, V. Suppose that

$$A_0 + A_1 + A_2 + A_3 = 0. \tag{7.20}$$

We get

$$|A_0A_1|^2 + |A_1A_2|^2 + |A_2A_3|^2 + |A_3A_0|^2 - 2(A_0^2 + A_1^2 + A_2^2 + A_3^2) =$$
$$(A_0 - A_1)^2 + (A_1 - A_2)^2 + (A_2 - A_3)^2 + (A_3 - A_0)^2 - 2(A_0^2 + A_1^2 + A_2^2 + A_3^2) =$$
$$-2(A_0 \cdot A_1 + A_1 \cdot A_2 + A_2 \cdot A_3 + A_3 \cdot A_0).$$

In view of (7.20) we have $A_0 + A_2 = -(A_1 + A_3)$ and

$$-2(A_0 \cdot A_1 + A_1 \cdot A_2 + A_2 \cdot A_3 + A_3 \cdot A_0) = -2(A_0 + A_2) \cdot (A_1 + A_3) = 2(A_1 + A_3)^2$$

which is greater than or equal to zero. From this Wirtinger's inequality (7.18) for $n = 4$ follows. The case of the equality is attained if and only if $A_1 = -A_3$, $A_0 = -A_2$, that is, for a parallelogram.

Remark 7.5.
1) The condition $\sum_{k=0}^{n-1} A_k = 0$ from Wirtinger's inequality (7.18) can always be satisfied if we put the origin of a coordinate system into the centroid T of an n-gon because of the relation $T = 1/n \sum_{k=0}^{n-1} A_k$ holding for the centroid of $A_0A_1 \ldots A_{n-1}$.

2) The inequality (7.18) is a discrete case of the continuous Wirtinger's inequality which is as follows [8, 54]:

Let $f(x)$ be an absolutely continuous real function with period 2π satisfying $\int_0^{2\pi} f(x)dx = 0$. Then

$$\int_0^{2\pi} f'(x)^2 dx \geq \int_0^{2\pi} f^2(x)dx. \tag{7.21}$$

The equality holds if and only if $f(x) = A\cos x + B\sin x$, where A, B are real constants.

7.4 Sum of squares inequality

In Chapter 6 we proved classically the inequality between the sum of squares of sides and the area of a triangle (6.8). We will call this *the sum of squares inequality.* Now we will prove (6.8) by computer. Let us recall the theorem [139]:

Theorem 7.10. *Given a triangle with side lengths a, b, c and area p. Then*

$$a^2 + b^2 + c^2 - 4\sqrt{3}\, p \geq 0. \tag{7.22}$$

The equality is attained if and only if a triangle is equilateral.

Introduce a Cartesian coordinate system so that $A = [x, y]$, $B = [0, 0]$, $C = [a, 0]$ (Fig. 7.9). We express the side lengths $a = |BC|$, $b = |AC|$,

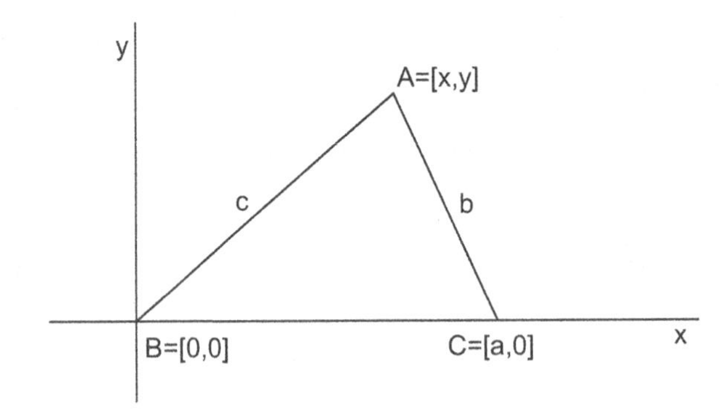

Fig. 7.9 Sum of squares inequality for a triangle: $a^2 + b^2 + c^2 - 4\sqrt{3}\, p \geq 0$

$c = |AB|$ and the area p in the form of algebraic equations:

$b = |AC| \Leftrightarrow h_1 : (x-a)^2 + y^2 - b^2 = 0,$

$c = |AB| \Leftrightarrow h_2 : x^2 + y^2 - c^2 = 0,$

$p =$ area of $ABC \Leftrightarrow h_3 : 2p - ay = 0.$

Further denote $q = \sqrt{3}$ which implies

$h_4 : q^2 - 3 = 0.$

Let k be a slack variable for which

$h_5 : a^2 + b^2 + c^2 - 4qp - k = 0.$

We want to express k in terms of *independent* variables x, y, k in such a form from which its non-negativity follows. In the ideal $I = (h_1, h_2, \ldots, h_5)$ we eliminate *dependent* variables b, c, p to obtain

```
Use R::=Q[xyabcpqk];
I:=Ideal((x-a)^2+y^2-b^2,x^2+y^2-c^2,2p-ay,q^2-3,
a^2+b^2+c^2-4qp-k);
Elim(b..p,I);
```

a few polynomials, one of which leads to the equation

$$-2yaq + 2x^2 + 2y^2 - 2xa + 2a^2 - k = 0. \tag{7.23}$$

From (7.23) we get

$$k = 2(x - a/2)^2 + 2(y - aq/2)^2. \tag{7.24}$$

We see that k is in the form of a sum of squares from which the validity of (7.22) follows. The equality in (7.22) is attained iff $x - a/2 = 0$ and $y - aq/2 = 0$, that is, the triangle ABC is equilateral. The theorem is proved.

Now consider the sum of squares inequality for quadrilaterals. It reads [8]:

Theorem 7.11. *Given a quadrilateral with side lengths a, b, c, d and area p. Then*

$$a^2 + b^2 + c^2 + d^2 - 4p \geq 0. \tag{7.25}$$

The equality holds if and only if the quadrilateral is a square.

We shall prove Theorem 7.11 by computer. Denote $a = |AB|$, $b = |BC|$, $c = |CD|$, $d = |CA|$ and let $A = [0,0]$, $B = [a,0]$, $C = [x,y]$, $D = [u,v]$ in a Cartesian coordinate system (Fig. 7.10). Then

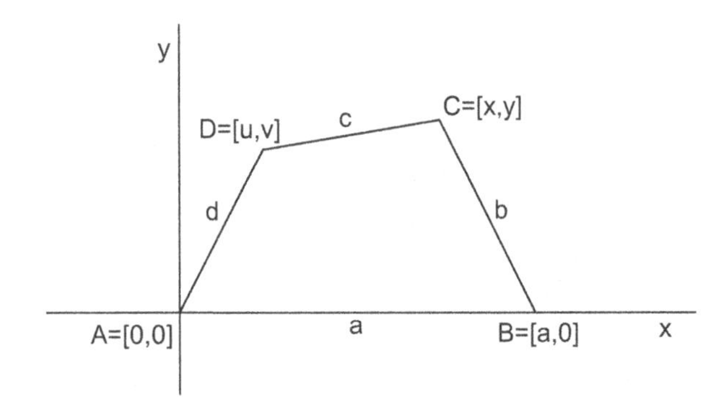

Fig. 7.10 Sum of squares inequality for a quadrilateral: $a^2+b^2+c^2+d^2-4p \geq 0$

$b = |BC| \Leftrightarrow h_1 : (x-a)^2 + y^2 - b^2 = 0,$

$c = |CD| \Leftrightarrow h_2 : (u-x)^2 + (v-y)^2 - c^2 = 0,$

$d = |DA| \Leftrightarrow h_3 : u^2 + v^2 - d^2 = 0,$

$p =$ area of $ABCD \Leftrightarrow h_4 : 2p - (ay + xv - yu) = 0.$

Denoting the difference on the left of (7.25) by k, we get

$h_5 : a^2 + b^2 + c^2 + d^2 - 4p - k = 0.$

In order to express k in terms of x, y, u, v, k we eliminate dependent variables b, c, d, p in the ideal $I = (h_1, h_2, \ldots, h_5)$ and obtain

```
Use R ::=Q[xyuvabcdpk];
I:=Ideal((x-a)^2+y^2-b^2,(u-x)^2+(v-y)^2-c^2,u^2+v^2-d^2,
2p-(ay+xv-yu),a^2+b^2+c^2+d^2-4p-k);
Elim(b..p,I);
```

the relation

$$k = 2(x^2 + y^2 - xu + yu + u^2 - xv - yv + v^2 - xa - ya + a^2)$$

which can be written in the form of a sum of squares as

$$k = (x-u-a)^2 + (y-a+u)^2 + (v-y)^2 + (v-x)^2. \quad (7.26)$$

We get the equality

$$a^2+b^2+c^2+d^2-4p = (x-u-a)^2+(y-a+u)^2+(v-y)^2+(v-x)^2 \quad (7.27)$$

from which we see that $k \geq 0$ with equality if and only if $v = x = y$ and $x - u - a = 0$ and $y - a + u = 0$. This implies $x = y - v = a, u = 0$ and the extremal quadrilateral is a square (Fig. 7.10). The theorem is proved.

Remark 7.6. The theorem on the sum of squares inequality for n-gons is as follows [8]:

Given an n-gon with the side lengths $a_1, a_2, \ldots, a_n$ and area p. Then

$$\sum_{i=1}^{n} a_i^2 - 4\, tan\frac{\pi}{n}\, p \geq 0, \tag{7.28}$$

with equality if and only if an n-gon is regular.

7.5 Isoperimetric inequality

Let us prove the following classical isoperimetric inequality for a triangle [15, 75]:

Theorem 7.12. *Let a, b, c be the side lengths of a triangle with area p. Then*

$$(a+b+c)^2 - 12\sqrt{3}\, p \geq 0. \tag{7.29}$$

The equality is attained if and only if a triangle is equilateral.

We shall prove the theorem by the sos (sum of squares) method. Let us introduce a Cartesian system of coordinates so that $A = [x, y]$, $B = [0, 0]$, $C = [a, 0]$ (Fig. 7.9). We have the following algebraic relations:

$b = |CA| \Leftrightarrow h_1 : (x-a)^2 + y^2 - b^2 = 0,$

$c = |AB| \Leftrightarrow h_2 : x^2 + y^2 - c^2 = 0,$

$p =$ area of $ABC \Leftrightarrow h_3 : 2p - ay = 0.$

Let $q = \sqrt{3}$ which implies

$h_4 : q^2 - 3 = 0,$

and let

$h_5 : (a+b+c)^2 - 12\sqrt{3}\, p - t = 0,$

where t is a slack variable. Eliminating x, y, p in the ideal $I = (h_1, h_2, \ldots, h_5)$ we express t in terms of the side lengths a, b, c. We get

```
Use R::=Q[xypqabct];
I:=Ideal((x-a)^2+y^2-b^2,x^2+y^2-c^2,2p-ay,q^2-3,
(a+b+c)^2-12qp-t);
Elim(x..q,I);
```

the equation of the form

$$t^2 + pt + q = 0, \tag{7.30}$$

where

$$p = -2(a+b+c)^2$$

and

$$q = 4(a+b+c)(7a^3-6a^2b-6ab^2+7b^3-6a^2c+15abc-6b^2c-6ac^2-6bc^2+7c^3).$$

We are to show that the roots t_1, t_2 of the quadratic equation (7.30) are non-negative. To prove this we shall use the formulas of Viète[1]

$$t_1 + t_2 \;=\; -p, \quad t_1 t_2 \;=\; q. \tag{7.31}$$

Thus by (7.31), it suffices to show that for non-negative numbers a, b, c, $p \leq 0$ and $q \geq 0$. Whereas the first inequality $p \leq 0$ is obvious, the second inequality $q \geq 0$ requires some knowledge of sos decomposition of q on the sum of squares of polynomials (see [21, 106] for details). It holds

$$q = 28[ab(a-b)^2+ac(a-c)^2+bc(b-c)^2]+14[(a-b)^2(a+b-c)^2+(a-c)^2(a-b+c)^2+(b-c)^2(-a+b+c)^2]+4(a+b+c)[a(b-c)^2+b(c-a)^2+c(a-b)^2]$$

from which non-negativity of q follows. This decomposition shows that $t = 0$ if and only if $a = b = c$, and the triangle ABC is equilateral.

The classical proof of the isoperimetric inequality (7.29) for a triangle is as follows [15]:
By the formula of Heron

$$16p^2 = (a+b+c)(-a+b+c)(a-b+c)(a+b-c). \tag{7.32}$$

The arithmetic-geometric inequality $\frac{x+y+z}{3} \;\geq\; (xyz)^{1/3}$ applied to $(x, y, z) = (-a+b+c, a-b+c, a+b-c)$ gives

$$\frac{(a+b+c)^3}{27} \geq (-a+b+c)(a-b+c)(a+b-c).$$

From here we get

$$(a+b+c)^4 \geq 27(a+b+c)(-a+b+c)(a-b+c)(a+b-c)$$

which implies (7.29) including the case of equality.

Remark 7.7. The isoperimetric inequality for n-gons is as follows [8, 75]:

Given an n-gon with side lengths $a_1, a_2, \ldots, a_n$ and area p. Then

$$\Big(\sum_{i=1}^{n} a_i\Big)^2 - 4n\, tan\frac{\pi}{n}\; p \;\geq\; 0, \tag{7.33}$$

with the equality only for a regular n-gon.

[1]In general case we use the Descartes' rule of signs [25].

7.6 Euler's inequality

In this part we will solve a geometric inequality by the method which differs from the previous one. This method is based on the expression of a given difference in terms of other variables which have a geometric meaning. The use of a coordinate system is not necessary.

In 1765, L. Euler [35] published the relation (7.34) which expresses the distance d of the incenter and circumcenter of an arbitrary triangle

$$d = \sqrt{r(r-2p)}\ , \tag{7.34}$$

where r is the circumradius and p is the inradius (Fig. 7.11).

H. Wieleitner [141] says that the relation (7.34) was published even earlier

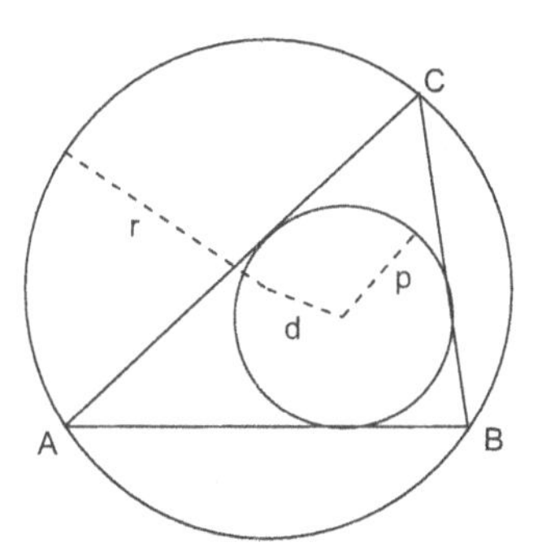

Fig. 7.11 Euler's relation: $d = \sqrt{r(r-2p)}$

by W. Chapple in 1746 (see [78]).

The following inequality

$$r \geq 2p\ , \tag{7.35}$$

which is a consequence of (7.34), is called *Euler's inequality* [17].

It is obvious that the equality in (7.35) is attained if and only if a triangle is equilateral, since only in this case the circumcenter and incenter coincide and $d = 0$ in (7.34).

First we will prove the Euler's inequality (7.35) by computer. We will proceed in a similar way as W. Koepf [65] without introducing a coordinate system.

Let ABC be an arbitrary triangle with side lengths a, b, c. Denote by r and p its circumradius and inradius respectively and let f be the area of ABC. For the area f of a triangle ABC we have the following well-known relations:

$$h_1 : f - p(a + b + c)/2 = 0,$$

$$h_2 : f - abc/(4r) = 0,$$

$$h_3 : 16f^2 - (a + b + c)(-a + b + c)(a - b + c)(a + b - c) = 0.$$

Suppose that a, b, c, p, r, f are positive real numbers. We will try to express $r - 2p$ in such a form from which it would be clear that $r - 2p$ is greater than or equal to zero. We introduce a slack variable k such that

$$h_4 : r - 2p - k = 0.$$

In the ideal $I = (h_1, h_2, h_3, h_4)$ we eliminate p and r to obtain polynomials in variables a, b, c, f, k. We enter

```
Use R::=Q[abcprfk];
I:=Ideal(2f-p(a+b+c),4fr-abc,16f^2-(a+b+c)(-a+b+c)(a-b+c)(a+b-
c),r-2p-k);
Elim(p..r,I);
```

and get a few polynomials from which the following one, after dividing it by a non-zero factor f, leads to the equation of the form

$$4fk = a^3 - a^2b - ab^2 + b^3 - a^2c + 3abc - b^2c - ac^2 - bc^2 + c^3. \quad (7.36)$$

From (7.36) we see that k is non-negative iff the polynomial on the right in (7.36) is non-negative.

To show non-negativity, we rearrange the polynomial on the right of (7.36) into the form (cf. [65]):

$$a^3 - a^2b - ab^2 + b^3 - a^2c + 3abc - b^2c - ac^2 - bc^2 + c^3$$

$$= 1/2[(a+b-c)(a-b)^2 + (b+c-a)(b-c)^2 + (c+a-b)(c-a)^2]. \quad (7.37)$$

The expression on the right-hand side in (7.37) is non-negative. Namely, it is the sum of non-negative expressions each of which consists of squares $(a - b)^2$, $(b - c)^2$, $(c - a)^2$ and expressions $a + b - c$, $b + c - a$, $c + a - b$ which are positive due to the triangle inequality.

The equality in (7.35) occurs if and only if $a = b = c$ in (7.37) on the right, and a triangle ABC is equilateral.

Remark 7.8. In order to prove (7.35) we could also prove the equality (7.34), from which the inequality (7.35) follows (see [136]).

Now we will add a classical proof of the Euler's inequality (7.35). It is ascribed to the Hungarian mathematician I. Ádám [78], whose proof is ingenious:

The midpoints A_1, B_1, C_1 of the sides of a triangle ABC form a triangle whose circumcircle has the radius $r/2$ (Fig. 7.12). Construct a triangle

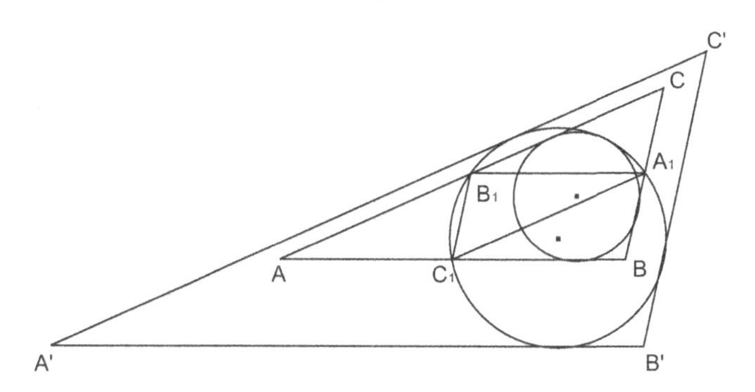

Fig. 7.12 Classical proof of the Euler's inequality

$A'B'C'$ whose sides are parallel to the sides of a triangle ABC so that the circumcircle of a triangle A_1, B_1, C_1 is the incircle of $A'B'C'$. Since $\triangle ABC \subseteq \triangle A'B'C'$, then for inradii $p, r/2$ of both triangles the relation $p \le r/2$ holds. The classical proof is now complete.

The proof just given can be easily applied to a tetrahedron and in general to an arbitrary simplex in n-dimensional Euclidean space E^n. For the radii p and r of spheres which are inscribed and circumscribed to a tetrahedron the relation

$$r \ge 3p , \tag{7.38}$$

which is analogous to (7.35), holds. The midpoints of sides are then replaced by the centroids of the faces of a tetrahedron.

Chapter 8

Regular polygons

We often encounter regular polygons in life — in civil engineering and architecture, in art, in animate and inanimate nature and in many other branches of human activities. Throughout history regular polygons played an important role, for instance, in proving the formula for the area and perimeter of a circle. The notion of a regular polygon is easily understandable. Thus it would seem that nothing new is possible to discover in this area. The following story is connected with one interesting discovery [132].

Before Christmas 1969 two chemists A. Dreiding and J. D. Dunitz visited the well-known mathematician B. L. van der Waerden. The latter talked about fixed and movable forms of cyclic hexane and octane. He also mentioned cyclic pentane and insisted that a skew pentagon with the same side lengths and the same angles must necessary lie in a plane. Van der Waerden was very astonished at this statement and asked for reasons.[1]

On 10th February 1970 at the Mathematics Colloquium in Zürich, B. L. van der Waerden held a lecture "Ein Satz über räumliche Fünfecke" whose only topic was a proof of the following theorem [132]:

A skew pentagon with equal sides and equal angles is planar.

After the publication of the paper [132] this theorem was proved in a short time by a number of mathematicians in a few ways, see [70, 127, 16, 57, 60, 133, 33, 69]. In [33] it is mentioned that this remarkable property of a pentagon was known in 1944 to J. Waser [137]. B. Grünbaum [47] writes that this property of a regular pentagon — i.e. equilateral and equiangular — was known to A. Auric [2] in 1911. Detailed information is described in the survey article [47], where it is stated that

[1] Also G. Pólya disclaimed any previous knowledge of the theorem and added "if Van der Waerden did not know about it then it was not known to mathematics", [33].

a total characterization of regular n-gons was done in 1962 by Russians V. A. Efremovitch and Ju. S. Iljjashenko [34].

In this section regular polygons in the Euclidean space E^d and their existence in spaces of various dimensions are studied from both the computer and classical perspectives. When proving that regular pentagons and heptagons span spaces of even dimension one encounters the case that the ideal I describing the hypotheses is not radical. Thus, in order to prove that $H \Rightarrow c$ one needs to show that c belongs to the radical of the ideal describing H. In practice we often encounter the case $\sqrt{I} = I$ which allows us to show that c is an element of the ideal I. S. Ch. Chou in his well-known book Mechanical Geometry Theorem Proving [22], where over five hundred examples are given, wrote on page 78 that " for all theorems we have found in practice $I = \sqrt{I}$ ". Exploring properties of regular polygons in E^d we meet the case in which it is necessary to show that the conclusion polynomial c belongs to the radical ideal $\sqrt{I}$. Regular polygons yield a suitable topic for investigating them from this point of view [93].

First we will give the definition of a regular polygon.

Definition 8.1. A polygon $P_0, P_1, \ldots, P_{n-1}$ whose sides have the same length, that is, $|P_jP_{j+1}|$ is constant for all $j = 0, 1, \ldots, n-1$, we call *equilateral*. Similarly, an n-gon is *k-equilateral* if $|P_jP_{j+\nu}| = d_\nu$ for all $j = 0, 1, \ldots, n-1$ and $\nu = 1, 2, \ldots, k$, where the constants d_ν are *parameters*. A polygon is called *regular* if for all $\nu = 1, 2, \ldots, n-1$ the lengths of segments $P_jP_{j+\nu}$ are independent of j, or in other words, if a polygon is $(n-1)$-equilateral.

Thus an equilateral n-gon is 1-equilateral with the parameter d_1, 2-equilateral n-gon is equilateral and equiangular with parameters d_1, d_2, etc. If we introduce $d_0 = 0$ then a regular n-gon is characterized by the relations

$$|P_jP_{j+\nu}| = d_\nu, \quad \text{for all} \quad j, \nu = 0, 1, \ldots, n-1, \tag{8.1}$$

which means that all diagonals of the "same" kind (next but one vertex, next but two vertices, ...) have the same length.

We say that an n-gon has the *dimension* s, if s is the dimension of the least subspace of the affine space A^d in which an n-gon is involved, or equivalently, if an n-gon spans an s-dimensional space.

8.1 Planarity of a regular pentagon

Consider the theorem [132]:

Theorem 8.1. *A regular skew pentagon ABCDE in the Euclidean space E^3 is given. Then ABCDE is a planar pentagon.*

First we shall prove the theorem by computer. Assume that a pentagon is equilateral with side length a and equiangular with diagonal length u.

Let us introduce a Cartesian coordinate system such that for the vertices of a pentagon $A = [a, 0, 0]$, $B = [b_1, b_2, 0]$, $C = [c_1, c_2, c_3]$, $D = [d_1, d_2, d_3]$, $E = [0, 0, 0]$ (Fig. 8.1). Then the following relations are fulfilled:

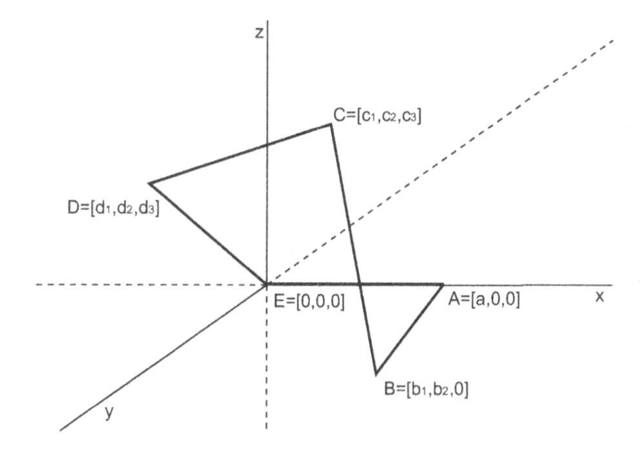

Fig. 8.1 Regular pentagon in E^3 is planar

$$
\begin{aligned}
&|AB| = a \Leftrightarrow h_1 : (b_1 - a)^2 + b_2^2 - a^2 = 0,\\
&|BC| = a \Leftrightarrow h_2 : (c_1 - b_1)^2 + (c_2 - b_2)^2 + c_3^2 - a^2 = 0,\\
&|CD| = a \Leftrightarrow h_3 : (d_1 - c_1)^2 + (d_2 - c_2)^2 + (d_3 - c_3)^2 - a^2 = 0,\\
&|DE| = a \Leftrightarrow h_4 : d_1^2 + d_2^2 + d_3^2 - a^2 = 0,\\
&|AC| = u \Leftrightarrow h_5 : (c_1 - a)^2 + c_2^2 + c_3^2 - u^2 = 0,\\
&|BD| = u \Leftrightarrow h_6 : (d_1 - b_1)^2 + (d_2 - b_2)^2 + d_3^2 - u^2 = 0,\\
&|CE| = u \Leftrightarrow h_7 : c_1^2 + c_2^2 + c_3^2 - u^2 = 0,\\
&|DA| = u \Leftrightarrow h_8 : (d_1 - a)^2 + d_2^2 + d_3^2 - u^2 = 0,\\
&|EB| = u \Leftrightarrow h_9 : b_1^2 + b_2^2 - u^2 = 0.
\end{aligned}
$$

The points A, B, C, D, E are complanar $\Leftrightarrow$ A, B, C, E and A, B, D, E are complanar $\Leftrightarrow$

$$z_1 : \begin{vmatrix} 0 & 0 & 0 & 1 \\ a & 0 & 0 & 1 \\ b_1 & b_2 & 0 & 1 \\ c_1 & c_2 & c_3 & 1 \end{vmatrix} = 0 \quad \text{and} \quad z_2 : \begin{vmatrix} 0 & 0 & 0 & 1 \\ a & 0 & 0 & 1 \\ b_1 & b_2 & 0 & 1 \\ d_1 & d_2 & d_3 & 1 \end{vmatrix} = 0. \tag{8.2}$$

We shall explore whether both conclusion polynomials z_1, z_2 belong to the radical of $I = (h_1, h_2, \dots, h_9)$. We will investigate each polynomial z_1, z_2 separately. For z_1 we find out whether $1 \in J$, where $J = I \cup \{ab_2c_3t - 1\}$. We enter

```
Use R::=Q[aub[1..3]c[1..3]d[1..3]t];
J:=Ideal((b[1]-a)^2+b[2]^2-a^2,(c[1]-b[1])^2+(c[2]-b[2])^2+
c[3]^2-a^2,(d[1]-c[1])^2+(d[2]-c[2])^2+(d[3]-c[3])^2-a^2,
d[1]^2+d[2]^2+d[3]^2-a^2,(c[1]-a)^2+c[2]^2+c[3]^2-u^2,(d[1]
-b[1])^2+(d[2]-b[2])^2+d[3]^2-u^2,c[1]^2+c[2]^2+c[3]^2-u^2,
(d[1]-a)^2+d[2]^2+d[3]^2-u^2,b[1]^2+b[2]^2-u^2,ab[2]c[3]t-1);
NF(1,J);
```

and get `NF(1,J)=0` which means that the points A, B, C, E are complanar. Similarly we will show that the points A, B, D, E are complanar as well. Whence we can conclude — a regular skew pentagon $ABCDE$ is planar.

In the last proof we examined the normal form `NF(1,J)`, where the ideal J contained the negated conclusion polynomial $ab_2c_3t - 1$. The result `NF(1,J)=0` means, that the conclusion polynomial ab_2c_3 is an element of the radical $\sqrt{I}$ from which $ab_2c_3 = 0$ follows. Usually it suffices to find out whether the conclusion polynomial ab_2c_3 belongs to the ideal I. Let us do it. We get

```
Use R::=Q[aub[1..3]c[1..3]d[1..3]st];
I:=Ideal((b[1]-a)^2+b[2]^2-a^2,(c[1]-b[1])^2+(c[2]-b[2])^2+
c[3]^2-a^2,(d[1]-c[1])^2+(d[2]-c[2])^2+(d[3]-c[3])^2-a^2,
d[1]^2+d[2]^2+d[3]^2-a^2,(c[1]-a)^2+c[2]^2+c[3]^2-u^2,(d[1]
-b[1])^2+(d[2]-b[2])^2+d[3]^2-u^2,c[1]^2+c[2]^2+c[3]^2-u^2,
(d[1]-a)^2+d[2]^2+d[3]^2-u^2,b[1]^2+b[2]^2-u^2);
NF(ab[2]c[3],I);
```

the non-zero result ab_2c_3. Hence the polynomial ab_2c_3 *does not* belong to the ideal I. However, ab_2c_3 belongs to the radical $\sqrt{I}$, i.e., there exists a natural number m such that $(ab_2c_3)^m$ belongs to the ideal I. It is easy to verify that $m = 3$, i.e., $(ab_2c_3)^3 \in I$.

Another proof of the fact that a regular pentagon in a space E^3 is planar, is based on the well-known theorem about the volume of a simplex expressed by the Cayley–Menger determinant (8.3). This kind of proof will be used later in the case of a regular heptagon. We apply the following formula for the volume of a simplex. Let us recall the definition of a simplex:

Definition 8.2. Points $A_1, A_2, \ldots, A_{n+1}$ of the affine space A^n form a *simplex* if the vectors $A_1 - A_2, A_1 - A_3, \ldots, A_1 - A_{n+1}$ are linearly independent.

For instance a segment is a simplex on the line, a triangle is a simplex in the plane, a tetrahedron is a simplex in the three dimensional space, etc.

Definition 8.3. The volume V_n of a simplex $A_1A_2 \ldots A_{n+1}$ in E^n can be expressed in terms of all mutual distances $|A_iA_j| = a_{ij}$ between vertices of a simplex in the form of the so-called Cayley–Menger's determinant. Denoting $a_{ij}^2 = d_{ij}$ then [5]:

$$(-1)^{n+1}2^n(n!)^2V_n^2 = D_n = \begin{vmatrix} 0 & 1 & 1 & 1 \ldots & 1 \\ 1 & 0 & d_{12} & d_{13} \ldots & d_{1,n+1} \\ 1 & d_{21} & 0 & d_{23} \ldots & d_{2,n+1} \\ \ldots & \ldots & \ldots & \ldots \ldots & \ldots \\ 1 & d_{n+1,1} & d_{n+1,2} & \ldots \ldots & 0 \end{vmatrix}. \tag{8.3}$$

We will prove Theorem 8.1 in the following generalized form [16]:

A regular pentagon $A_1A_2A_3A_4A_5$ lies either in E^4 or in E^2, i.e., its dimension is 4 or 2.

The proof, which was published by O. Bottema [16] in 1973 (see also [60]), is as follows.

If we put $V_n = 0$ then by (8.3) also $D_n = 0$ and points $A_1, A_2, \ldots, A_{n+1}$ span a space whose dimension is less than n.

Minimal dimension of a space in which an arbitrary pentagon can be considered is four. Let $A_1A_2A_3A_4A_5$ be a regular pentagon in the Euclidean space E^4. The vertices of a pentagon in E^4 form a simplex whose volume V_4 can be expressed in terms of all distances between the vertices of a simplex. Denote the lengths of its sides and diagonals by

$$|A_1A_2| = |A_2A_3| = |A_3A_4| = |A_4A_5| = |A_5A_1| = a,$$

$$|A_1A_3| = |A_3A_5| = |A_5A_2| = |A_2A_4| = |A_4A_1| = b.$$

By (8.3) for a simplex $A_1A_2A_3A_4A_5$ in E^4 it holds

$$D_4 = \begin{vmatrix} 0 & 1 & 1 & 1 & 1 & 1 \\ 1 & 0 & a^2 & b^2 & b^2 & a^2 \\ 1 & a^2 & 0 & a^2 & b^2 & b^2 \\ 1 & b^2 & a^2 & 0 & a^2 & b^2 \\ 1 & b^2 & b^2 & a^2 & 0 & a^2 \\ 1 & a^2 & b^2 & b^2 & a^2 & 0 \end{vmatrix} = -5(a^2+ab-b^2)^2(a^2-ab-b^2)^2. \tag{8.4}$$

We can compute the determinant (8.4) directly by hand or we can take advantage of the command `Det(M)` in CoCoA which assigns to a matrix M its determinant. We enter

```
Use R::=Q[ab];
M:=Mat([[0,1,1,1,1,1],[1,0,a^2,b^2,b^2,a^2],[1,a^2,0,a^2,b^2,
b^2],[1,b^2,a^2,0,a^2,b^2],[1,b^2,b^2,a^2,0,a^2],[1,a^2,b^2,
b^2,a^2,0]]);
Det(M);
```

and get the polynomial $-5a^8+30a^6b^2-55a^4b^4+30a^2b^6-5b^8$, from which we obtain, using the command `Factor`, the result above (8.4).

Suppose that $D_4 = 0$. This means that the dimension of the pentagon $A_1A_2A_3A_4A_5$ is less than or equal to three.

Now consider four vertices of $A_1A_2A_3A_4A_5$, for instance A_1, A_2, A_3, A_4, which form a simplex in the space E^3 and explore its volume. By (8.3) we omit the last row and column in the determinant (8.4). We get

$$D_3 = \begin{vmatrix} 0 & 1 & 1 & 1 & 1 \\ 1 & 0 & a^2 & b^2 & b^2 \\ 1 & a^2 & 0 & a^2 & b^2 \\ 1 & b^2 & a^2 & 0 & a^2 \\ 1 & b^2 & b^2 & a^2 & 0 \end{vmatrix} = -2(a^2+ab-b^2)(a^2-ab-b^2)(a^2+b^2). \tag{8.5}$$

Hence

$$D_4 = 0 \qquad \Rightarrow \qquad D_3 = 0. \tag{8.6}$$

In order to prove (8.6) algebraically, we necessarily need to show that D_3 belongs to the radical ideal $\sqrt{D_4}$ since the polynomial D_3 *is not* an element of the ideal generated by D_4. We can see this without the use of a computer. We obtain the same result for remaining quadruples of the vertices of a pentagon.

Thus, if the dimension of the pentagon $A_1A_2A_3A_4A_5$ is not 4 then it is neither 3. Therefore the pentagon $A_1A_2A_3A_4A_5$ must be planar. The theorem is proved.

Remark 8.1. Realize that we are working in the theory where polynomials have variables in *complex* numbers. From this point of view the relation (8.6), namely $D_4 = 0 \Rightarrow D_3 = 0$, is correct. But in *real* numbers the equivalence $D_4 = 0 \Leftrightarrow D_3 = 0$ holds.

Let us show a classical proof of the statement that an equilateral and equiangular pentagon in E^3 is necessarily planar, which is due to G. Bol and H. S. M. Coxeter [133, 47]. This proof is short and elegant:

Given side length a and angle α, all mutual distances of vertices of a regular pentagon $A_1A_2A_3A_4A_5$ are determined. Then there exists a congruent mapping $\mathcal{S}$ which cyclically permutes the vertices, i.e., it assigns to a pair A_iA_j the pair $A_{i+1}A_{j+1}$, for $i, j = 1, 2, \ldots, 5$. Since composite mapping $\mathcal{S}^5$ is an identity, the congruent mapping $\mathcal{S}$ cannot be a plane symmetry. Further, the mapping $\mathcal{S}$ has a fixed point — the centroid of a pentagon, therefore we can rule out a translation. Thus $\mathcal{S}$ is a rotation and a pentagon lies in the plane which is orthogonal to the axis of rotation.

A regular pentagon $A_1A_2A_3A_4A_5$ in a plane with the length of sides a and diagonals b obeys by (8.4) and (8.5) the equation

$$(a^2 + ab - b^2)(a^2 - ab - b^2) = 0. \tag{8.7}$$

The equation

$$a^2 + ab - b^2 = 0 \tag{8.8}$$

holds for a convex regular pentagon (Fig. 8.2 on the left), from which

$$\frac{a}{b} = \frac{-1 + \sqrt{5}}{2} = 0.618...$$

follows.

From $a^2 - ab - b^2 = 0$ we get the relation

$$\frac{a}{b} = \frac{1 + \sqrt{5}}{2} = 1.618...$$

which characterizes non-convex star regular pentagons (sometimes we call it a pentagram) (Fig. 8.2 on the right).

The number $\Phi = \frac{-1+\sqrt{5}}{2} = 0.618...$ is called the *golden ratio*. It occurs both in nature and in many branches of human activity. We say that

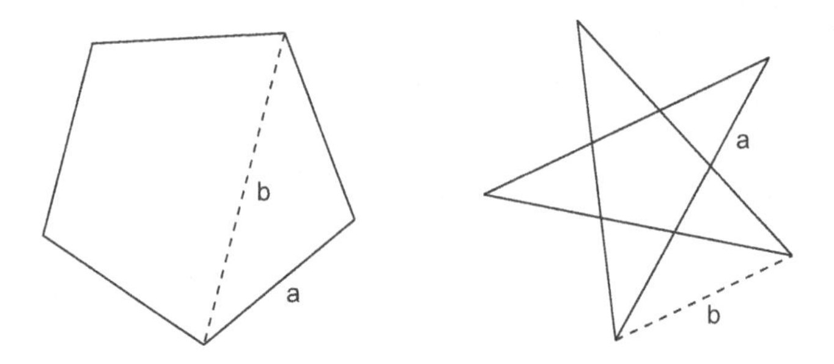

Fig. 8.2 Convex and non-convex regular pentagons

a segment which is divided into two parts in the ratio Φ, is divided in the golden ratio.

We can derive the equation (8.8) also directly, for instance in this way. In a convex regular pentagon it holds (Fig. 8.2 on the left),

$$a = 2R\sin\frac{\pi}{5}, \quad b = 2R\sin\frac{2\pi}{5}, \tag{8.9}$$

where R is the circumradius of a pentagon. We come out from the identity

$$\sin\frac{2\pi}{5} = \sin\frac{3\pi}{5}$$

which is obvious. Multiplication of this equation by $\sin\frac{2\pi}{5}$ with the use of the double angle formula gives

$$\sin^2\frac{2\pi}{5} = 2\sin\frac{\pi}{5}\cos\frac{\pi}{5}\sin\frac{3\pi}{5}.$$

Using the product formula for the trigonometric functions we obtain

$$\sin^2\frac{2\pi}{5} = \sin\frac{\pi}{5}(\sin\frac{4\pi}{5} + \sin\frac{2\pi}{5}).$$

From here, with help of (8.9), we get (8.8).

Similarly, the relation for a non-convex regular pentagon can be proved. It suffices to exchange a and b in (8.8) (Fig. 8.2 on the right). From (8.4) and (8.5) we get the theorem [16]:

A regular pentagon with the lengths of sides a and diagonals b is planar if and only if condition (8.7) holds. The equation $a^2 + ab - b^2 = 0$ gives a convex regular pentagon, whereas $a^2 - ab - b^2 = 0$ leads to a non-convex regular pentagon.

8.2 Regular heptagon

The case of a regular pentagon from the previous part is well-known and has been published many times. On the other hand a regular heptagon and its existence in spaces of various dimensions has not been mentioned explicitly so many times as a regular pentagon. This is likely due to the following general theorem on regular polygons with an arbitrary number of vertices [9, 34, 47].

A regular polygon with an odd number of vertices has even dimension.

Let us look at a regular heptagon in detail. We will investigate its properties using the method which is based on Gröbner bases computation. We will apply the method which was used by a regular pentagon and which is based on the Cayley–Menger determinant.

Let $A_1A_2A_3A_4A_5A_6A_7$ be a regular heptagon. In accordance with the definition 8.1 a regular heptagon is 3-equilateral. Here we have to pay attention to the fact that the conditions being equilateral and equiangular, which were sufficient for a pentagon to be regular, do not suffice for a regular heptagon. For instance in Fig. 8.3 we see an equilateral and equiangular

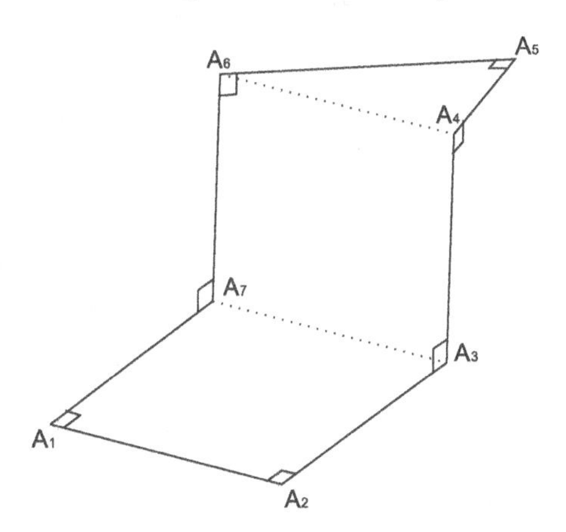

Fig. 8.3 A skew equilateral and equiangular heptagon which is not regular

heptagon which is not regular since $|A_7A_3| \neq |A_1A_4|$ (see [47]).

In a regular heptagon $A_1A_2A_3A_4A_5A_6A_7$ let us denote

$$|A_1A_2| = |A_2A_3| = |A_3A_4| = |A_4A_5| = |A_5A_6| = |A_6A_7| = |A_7A_1| = a,$$

$$|A_1A_3| = |A_3A_5| = |A_5A_7| = |A_7A_2| = |A_2A_4| = |A_4A_6| = |A_6A_1| = b,$$

$$|A_1A_4| = |A_4A_7| = |A_7A_3| = |A_3A_6| = |A_6A_2| = |A_2A_5| = |A_5A_1| = c.$$

Consider a heptagon in a six dimensional Euclidean space E^6 which is a space of minimal dimension in which an arbitrary heptagon can be placed. By the formula (8.3) for the volume of a simplex $A_1A_2A_3A_4A_5A_6A_7$ in E^6 it holds

$$D_6 = \begin{vmatrix} 0 & 1 & 1 & 1 & 1 & 1 & 1 & 1 \\ 1 & 0 & a^2 & b^2 & c^2 & c^2 & b^2 & a^2 \\ 1 & a^2 & 0 & a^2 & b^2 & c^2 & c^2 & b^2 \\ 1 & b^2 & a^2 & 0 & a^2 & b^2 & c^2 & c^2 \\ 1 & c^2 & b^2 & a^2 & 0 & a^2 & b^2 & c^2 \\ 1 & c^2 & c^2 & b^2 & a^2 & 0 & a^2 & b^2 \\ 1 & b^2 & c^2 & c^2 & b^2 & a^2 & 0 & a^2 \\ 1 & a^2 & b^2 & c^2 & c^2 & b^2 & a^2 & 0 \end{vmatrix} \qquad (8.10)$$

$$= -7(a^6+3a^4b^2-4a^2b^4+b^6-4a^4c^2-a^2b^2c^2+3b^4c^2+3a^2c^4-4b^2c^4+c^6)^2.$$

Now we express the volume of a simplex $A_1A_2A_3A_4A_5A_6$ in E^5 deleting the last row and column in the previous determinant (8.10). The result is

$D_5 = 2(2a^4 - 3a^2b^2 + 2b^4 - 3a^2c^2 - 3b^2c^2 + 2c^4)(a^6 + 3a^4b^2 - 4a^2b^4 + b^6 - 4a^4c^2 - a^2b^2c^2 + 3b^4c^2 + 3a^2c^4 - 4b^2c^4 + c^6).$

The same result we obtain for other 6-tuples of the vertices of a heptagon. The comparison of the determinants D_6 and D_5 gives

$$D_6 = 0 \qquad \Rightarrow \qquad D_5 = 0. \qquad (8.11)$$

To prove (8.11) algebraically, we immediately see that the polynomial D_5 is an element of the radical $\sqrt{D_6}$. Note that D_5 *does not* belong to the ideal which is generated by D_6.

Hence, if the dimension of a regular heptagon is not six then it is neither five. Therefore the heptagon must lie in the space of dimension four or less.

Suppose that $D_4 = 0$ for all 5-tuples of the vertices of a heptagon. It is obvious that (see Fig. 8.4) it suffices to explore three simplices $A_1A_2A_3A_4A_5$, $A_1A_2A_3A_4A_6$, and $A_1A_2A_3A_5A_6$ in E^4. For a simplex $A_1A_2A_3A_4A_5$ we get

$$D_4(12345) = -(a^2-bc-c^2)(a^2+bc-c^2)(a^4-12a^2b^2+8b^4+2a^2c^2-5b^2c^2+c^4)$$

and analogously

$D_4(12346) = -(a^2-b^2-ac)(a^2-b^2+ac)(a^4+2a^2b^2+b^4-5a^2c^2-12b^2c^2+8c^4),$

$$D_4(12356) = (ab-b^2+c^2)(ab+b^2-c^2)(8a^4-5a^2b^2+b^4-12a^2c^2+2b^2c^2+c^4).$$

The condition $D_4 = 0$, i.e. the validity of the equations

$$D_4(12345) = 0, \ D_4(12346) = 0, \ D_4(12356) = 0,$$

means that a heptagon $A_1A_2A_3A_4A_5A_6A_7$ is of dimension three or less. Investigating the volume of a tetrahedron $A_1A_2A_3A_4$ in E^3 we obtain

$$D_3(1234) = \begin{vmatrix} 0 & 1 & 1 & 1 & 1 \\ 1 & 0 & a^2 & b^2 & c^2 \\ 1 & a^2 & 0 & a^2 & b^2 \\ 1 & b^2 & a^2 & 0 & a^2 \\ 1 & c^2 & b^2 & a^2 & 0 \end{vmatrix} = -2(a^2-b^2+ac)(a^2-b^2-ac)(a^2+2b^2-c^2).$$

Similarly, considering all quadruples of the vertices of a heptagon, we get another three results:

$$D_3(1235) = -2(a^4b^2 - 5a^2b^2c^2 + b^4c^2 + a^2c^4),$$

$$D_3(1245) = -2(a^2 - bc - c^2)(2a^2 - b^2 + c^2)(a^2 + bc - c^2),$$

$$D_3(1246) = -2(a^2 - b^2 - 2c^2)(ab + b^2 - c^2)(ab - b^2 + c^2).$$

We will prove that

$$D_4 = 0 \qquad \Rightarrow \qquad D_3 = 0, \tag{8.12}$$

that is, from the hypotheses

$$D_4(12345) = 0, \ D_4(12346) = 0, \ D_4(12356) = 0$$

the conclusions

$$D_3(1234) = 0, \ D_3(1235) = 0, \ D_3(1245) = 0, \ D_3(1246) = 0$$

follow.

Let us denote by $I = (D_4(12345), D_4(12346), D_4(12356))$, and $J = (D_3(1234), D_3(1235), D_3(1245), D_3(1246))$ the respective ideals and compute their radicals. In CoCoA we enter

```
Use R::=Q[abc];
I:=Ideal(-(a^2-bc-c^2)(a^2+bc-c^2)(a^4-12a^2b^2+8b^4+2a^2c^2-
5b^2c^2+c^4),-(a^2-b^2-ac)(a^2-b^2+ac)(a^4+2a^2b^2+b^4-5a^2c^2
-12b^2c^2+8c^4),(ab-b^2+c^2)(ab+b^2-c^2)(8a^4-5a^2b^2+b^4-
12a^2c^2+2b^2c^2+c^4));
Radical(I);
```

and get

$$\sqrt{I} = (a^4 - 2a^2c^2 - b^2c^2 + c^4, a^2b^2 - a^2c^2 - 3b^2c^2 + 2c^4, b^4 - a^2c^2 - 5b^2c^2 + 3c^4). \tag{8.13}$$

We acquire the same result for $\sqrt{J}$. Hence, the radicals of both ideals I and J are alike.

Thus we proved even more then (8.12), namely that $D_4 = 0 \Leftrightarrow D_3 = 0$. Whence, if the dimension of a regular heptagon is not four then it is neither three. Therefore a heptagon $A_1A_2A_3A_4A_5A_6A_7$ must lie in a plane. We proved the theorem [9, 34]:

Theorem 8.2. *A regular heptagon lies either in the Euclidean space E^6 or in E^4 or in E^2, i.e. its dimension is either* 6 *or* 4 *or* 2.

Remark 8.2. If we apply the weaker criterion in the theorem above to investigate whether the polynomial $D_3(1234)$ is an element of the ideal I instead of its radical $\sqrt{I}$, we get

```
Use R::=Q[abc];
I:=Ideal(-(a^2-bc-c^2)(a^2+bc-c^2)(a^4-12a^2b^2+8b^4+2a^2c^2-
5b^2c^2+c^4),-(a^2-b^2-ac)(a^2-b^2+ac)(a^4+2a^2b^2+b^4-5a^2c^2
-12b^2c^2+8c^4),(ab-b^2+c^2)(ab+b^2-c^2)(8a^4-5a^2b^2+b^4-
12a^2c^2+2b^2c^2+c^4));
NF(-2(a^2-b^2+ac)(a^2-b^2-ac)(a^2+2b^2-c^2),I);
```

a non-zero result $-2a^6 + 6a^2b^4 - 4b^6 + 4a^4c^2 + 2b^4c^2 - 2a^2c^4$. Hence the polynomial

$$D_3(1234) = -2(a^2 - b^2 + ac)(a^2 - b^2 - ac)(a^2 + 2b^2 - c^2)$$

is not an element of the ideal I. We again encounter the case in which it is necessary to examine the membership of the conclusion polynomial $D_3(1234)$ in the radical $\sqrt{I}$. A close inspection shows that the polynomial $(D_3(1234))^3$ belongs to the ideal I.

The results above enable us to characterize regular heptagons lying in a plane. A theorem holds:

Theorem 8.3. *A regular heptagon $A_1A_2\ldots A_7$ in E^d with lengths of sides and diagonals $a = |A_iA_{i+1}|$, $b = |A_iA_{i+2}|$, $c = |A_iA_{i+3}|$, where $i = 1, 2, \ldots, n-1$, is planar if and only if the following equations hold*

$$\begin{aligned} h_1 &: (a^2 - bc - c^2)(a^2 + bc - c^2) = 0, \\ h_2 &: (a^2 - b^2 - ac)(a^2 - b^2 + ac) = 0, \\ h_3 &: (ab - b^2 + c^2)(ab + b^2 - c^2) = 0. \end{aligned} \tag{8.14}$$

Proof. First suppose that a regular heptagon $A_1A_2\dots A_7$ is planar. Let us prove for instance the first equation in (8.14). We will show that the polynomial h_1 belongs to the radical $\sqrt{I}$ (8.13), where $I = (D_4(12345), D_4(12346), D_4(12356))$. Computation in CoCoA gives

```
Use R::=Q[abc];
I:=Ideal(-(a^2-bc-c^2)(a^2+bc-c^2)(a^4-12a^2b^2+8b^4+2a^2c^2-
5b^2c^2+c^4),-(a^2-b^2-ac)(a^2-b^2+ac)(a^4+2a^2b^2+b^4-5a^2c^2
-12b^2c^2+8c^4),(ab-b^2+c^2)(ab+b^2-c^2)(8a^4-5a^2b^2+b^4-
12a^2c^2+2b^2c^2+c^4));
NF((a^2-bc-c^2)(a^2+bc-c^2),Radical(I));
```

the result 0. Similarly we prove other equations in (8.14).

Conversely, suppose that equations (8.14) hold. We are to prove that $D_4 = 0$, that is, $D_4(12345) = D_4(12346) = D_4(12356) = 0$. Let us prove that $D_4(12346) = 0$. We will show that the polynomial $D_4(12346)$ belongs to the ideal $K = (h_1, h_2, h_3)$. The normal form of $D_4(12346)$ with respect to the Gröbner basis of the ideal K

```
Use R::=Q[abc];
K:=Ideal((a^2-bc-c^2)(a^2+bc-c^2),(a^2-b^2-ac)(a^2-b^2+ac),
(ab-b^2+c^2)(ab+b^2-c^2));
NF(-(a^2-b^2-ac)(a^2-b^2+ac)(a^4+2a^2b^2+b^4-5a^2c^2-12b^2c^2
+8c^4),K);
```

equals 0. Similarly we show that $D_4(12345) = 0$ and $D_4(12356) = 0$. The theorem is proved. □

Remark 8.3.

1) If $K = (h_1, h_2, h_3)$ is the ideal generated by the polynomials on the left sides of (8.14) then $\sqrt{I} = \sqrt{K}$. Hence, the ideals I and K have the same radicals. From this Theorem 8.3 follows.

2) To prove that the polynomials h_1, h_2, h_3 on the left sides of (8.14) belong to the ideal I, fails. All these polynomials belong to the radical $\sqrt{I}$. It is easy to show that h_1^4, h_2^4 and h_3^4 belong to the ideal I.

By Theorem 8.3 we can compute lengths of sides and diagonals a, b, c of a regular heptagon in E^2. The system of equations

$$a^2 + bc - c^2 = 0, \qquad a^2 - b^2 + ac = 0 \tag{8.15}$$

characterizes a convex regular heptagon (Fig. 8.4).

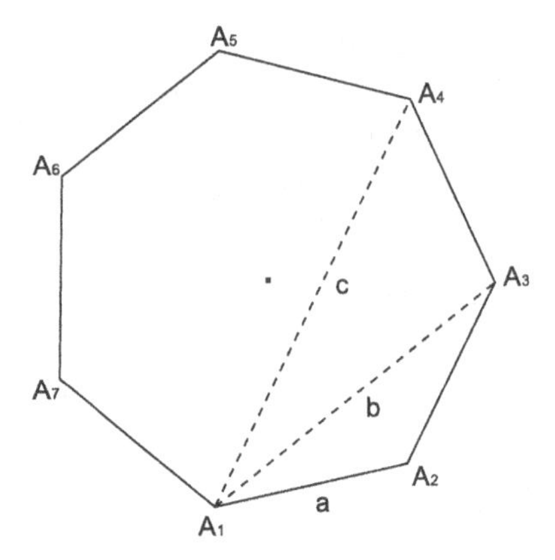

Fig. 8.4 Convex regular heptagon

Similarly, we shall describe the equations for the remaining two non-convex star regular heptagons (Fig. 8.5). For a heptagon in Fig. 8.5 on the left (so-called 2-regular heptagon) it holds

$$a^2 - bc - c^2 = 0, \qquad ab - b^2 + c^2 = 0, \tag{8.16}$$

and for a heptagon in Fig. 8.5 on the right (so called 3-regular heptagon)

$$a^2 - b^2 - ac = 0, \qquad ab + b^2 - c^2 = 0. \tag{8.17}$$

Let us prove directly the first equation in (8.15)

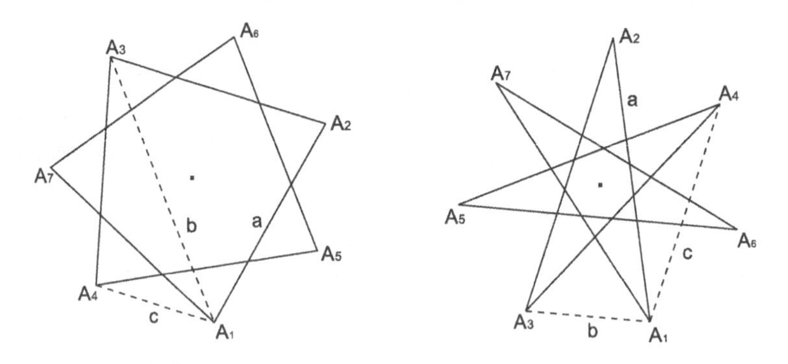

Fig. 8.5 Non-convex 2-regular and 3-regular heptagons

$$a^2 + ac - b^2 = 0, \tag{8.18}$$

which holds for a convex regular heptagon [126] (Fig. 8.4). Obviously

$$a = 2R\sin\frac{\pi}{7}\,, \quad b = 2R\sin\frac{2\pi}{7}\,, \quad c = 2R\sin\frac{3\pi}{7}\,, \tag{8.19}$$

where R is the circumradius of a heptagon. We will use the identity

$$\sin\frac{2\pi}{7} = \sin\frac{5\pi}{7}\,.$$

Multiplication of this equation by $\sin\frac{2\pi}{7}$ together with the use of the double angle formula yields

$$\sin^2\frac{2\pi}{7} = 2\sin\frac{\pi}{7}\cos\frac{\pi}{7}\sin\frac{5\pi}{7}\,.$$

The use of the product formula for trigonometric functions gives

$$\sin^2\frac{2\pi}{7} = \sin\frac{\pi}{7}(\sin\frac{6\pi}{7} + \sin\frac{4\pi}{7})\,.$$

Multiplying this equation by $4R^2$ with respect to (8.19), we obtain (8.18).

Relations (8.15), (8.16), (8.17) can also be proved by the Ptolemy theorem (Fig. 8.4). Consider the cyclic quadrilateral $A_1A_2A_3A_4$. Then by Ptolemy's theorem $a^2 + ac - b^2 = 0$. Analogously from the cyclic quadrilateral $A_1A_2A_4A_5$ we get $a^2 + bc - c^2 = 0$. These are relations in (8.15). Notice that using the quadrilateral $A_1A_2A_3A_5$ we obtain another relation $ab + ac - bc = 0$ which characterizes a convex regular heptagon in a plane.

Regular polygons play an important role in the isoperimetric inequality. The isoperimetric inequality for n-gons in a plane reads [8]:

From all n-gons in a plane of a given perimeter, a regular n-gon has the greatest area.

An analogue of the isoperimetric inequality in a space is as follows [11]:

From all skew n-gons in E^d with the given perimeter find an n-gon of the maximal volume of its convex hull.

This spatial analogue of the isoperimetric inequality for n-gons was successfully solved in spaces of *even* dimension d, where extremal n-gons are more-dimensional analogues of regular polygons [83, 116].

In a space of *odd* dimension d this problem has not been solved yet. A few special cases in E^3 for a skew quadrilateral [76], pentagon and hexagon [63, 64] have been solved. Especially an extremal heptagon in E^3 is of interest since by the theory above it *can not* be regular. Hence a natural question arises, what do extremal n-gons in E^3 for $n \geq 7$ look like?

Chapter 9

Miscellaneous

In this chapter we will give a few problems from various parts of geometry which seem to the author to be of interest. Especially the first two parts — "Non-elementary constructions" and "Loci of points of given properties" — are important topics both from the geometry and algebra point of view. Whereas loci of points belong to well-known issues in the theory of elimination (see for example the book [136] by D. Wang), non-elementary constructions are not often mentioned in computer algebra topics despite the fact that nowadays we can use computer algebra tools to make constructions which were infeasible in the past.

Further we will be concerned with the theorem of Viviani and the theorem on a complete quadrilateral. When proving the theorem of Viviani we used the method of automatic discovery based on the extension of a conclusion variety which leads to the generalization of this classical result. The theorem on Gauss' line (or also the theorem on a complete quadrilateral) is investigated both by computer and classically. Comparing these two approaches shows that sometimes one method seems to be easier than the other (in this case the former).

At the end of this chapter most common problems that the students encounter during the seminar of automatic theorem proving which the author held at the university for several years are given. The experience that I had obtained during the seminar are included as well.

9.1 Non-elementary constructions

Due to modern tools like computers and mathematical software and the theory of elimination we are able to carry out even non-elementary constructions. By a non-elementary construction we mean a construction which is either difficult to construct by compass and ruler or which is not possible to construct using these tools. Non-elementary constructions can be divided into two groups — algebraic or transcendental — according to the equations that we get using an analytic method to describe the problem. An example of such a non-elementary construction is the following problem (see [59]):

Given four lines a, b, c, d in the plane, construct a square $KLMN$ such that $K \in a$, $L \in b$, $M \in c$, $N \in d$.

First we will use a computer to solve the problem. Let us choose a Cartesian coordinate system so that the vertices K, L, M, N of a square have coordinates $K = [k_1, k_2], L = [l_1, l_2], M = [m_1, m_2], N = [n_1, n_2]$ (Fig. 9.1). Let the lines a, b, c, d be defined by the equations:

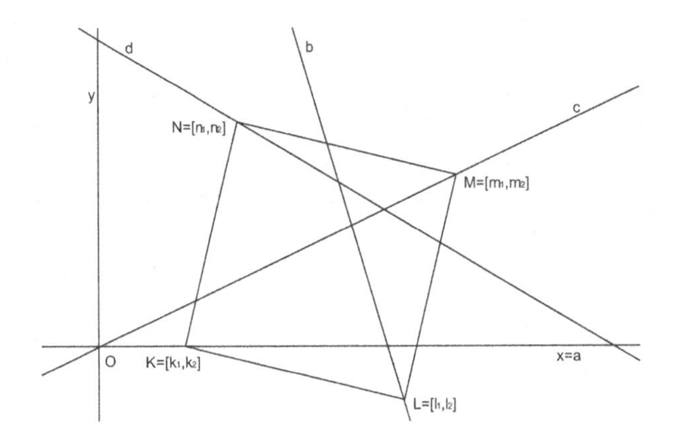

Fig. 9.1 Square $KLMN$ whose vertices lie on the given lines a, b, c, d.

$a : a_1x + a_2y + a_3 = 0,\ b : b_1x + b_2y + b_3 = 0,\ c : c_1x + c_2y + c_3 = 0,$ $d : d_1x + d_2y + d_3 = 0.$

Then

$$K \in a \Leftrightarrow h_1 : a_1k_1 + a_2k_2 + a_3 = 0,$$

$$L \in b \Leftrightarrow h_2 : b_1l_1 + b_2l_2 + b_3 = 0,$$

$$M \in c \Leftrightarrow h_3 : c_1m_1 + c_2m_2 + c_3 = 0,$$

$$N \in d \Leftrightarrow h_4 : d_1n_1 + d_2n_2 + d_3 = 0.$$

To ensure that $KLMN$ is a square, whose vertices K, L, M, N lie on the lines a, b, c, d respectively, we rotate the vector $L - K$ by 90° in a positive sense. We get the vector $N - K$ which we denote by $\text{rot}(L - K) = N - K$. Then we will rotate the vector $K - N$ by 90° in the same sense and get the vector $M - N$, etc.
We have the following relations:

$\text{rot}(L - K) = N - K \Leftrightarrow$
$h_5 : -(l_2 - k_2) - (n_1 - k_1) = 0,\ h_6 : l_1 - k_1 - (n_2 - k_2) = 0,$

$\text{rot}(K - N) = M - N \Leftrightarrow$
$h_7 : -(k_2 - n_2) - (m_1 - n_1) = 0,\ h_8 : k_1 - n_1 - (m_2 - n_2) = 0,$

$\text{rot}(N - M) = L - M \Leftrightarrow$
$h_9 : -(n_2 - m_2) - (l_1 - m_1) = 0,\ h_{10} : n_1 - m_1 - (l_2 - m_2) = 0,$

$\text{rot}(M - L) = K - L \Leftrightarrow$
$h_{11} : -(m_2 - l_2) - (k_1 - l_1) = 0,\ h_{12} : m_1 - l_1 - (k_2 - l_2) = 0.$

We get the system of 12 equations with 8 unknowns k_1, k_2, l_1, l_2, m_1, m_2, n_1, n_2. There is no loss of generality if we put $a_1 = 0$, $a_2 = 1$, $a_3 = 0$, $c_3 = 0$ (Fig. 9.1). In the ideal $I = (h_1, h_2, \ldots, h_{12})$ we eliminate dependent variables $k_2, \ldots, n_2$ except k_1 and get

```
Use R::=Q[a[1..3]b[1..3]c[1..3]d[1..3]k[1..2]l[1..2]m[1..2]
n[1..2]];
I:=Ideal(k[2],b[1]l[1]+b[2]l[2]+b[3],c[1]m[1]+c[2]m[2],d[1]n[1]
+d[2]n[2]+d[3],-(l[2]-k[2])-(n[1]-k[1]),l[1]-k[1]-(n[2]-k[2]),
-(k[2]-n[2])-(m[1]-n[1]),k[1]-n[1]-(m[2]-n[2]),-(n[2]-m[2])-
(l[1]-m[1]),n[1]-m[1]-(l[2]-m[2]),-(m[2]-l[2])-(k[1]-l[1]),m[1]
-l[1]-(k[2]-l[2]),a[1],a[2]-1,a[3],c[3]);
Elim(k[2]..n[2],I);
```

the result

$$k_1 = \frac{-b_3c_1d_1 - b_3c_2d_1 + b_3c_1d_2 - b_3c_2d_2 - b_1c_1d_3 - b_2c_1d_3 + b_1c_2d_3 - b_2c_2d_3}{b_1c_1d_1 + b_2c_1d_1 + b_2c_2d_1 - b_1c_1d_2 - b_2c_1d_2 + b_1c_2d_2}.$$

Similarly we find the remaining unknowns. We can see one of solutions in Fig. 9.1. From the construction it is seen that there exist at most four squares $KLMN$ with the given properties.

We can also proceed in a different way as A. Karger [59].
Denote the coordinates of the vertices of a square $KLMN$ and equations of the straight lines a, b, c, d as above and consider a square $ABCD$ with

the vertices $A = [1, 0], B = [0, 1], C = [-1, 0], D = [0, -1]$. We search for an equiform transformation φ

$$\varphi : x' = px - qy + r, \; y' = qx + py + s, \tag{9.1}$$

where p, q, r, s are unknown real coefficients, which transforms the square $ABCD$ into $KLMN$. From (9.1) we get

$\varphi([1, 0]) = [p + r, q + s] = [k_1, k_2],$

$\varphi([0, 1]) = [-q + r, p + s] = [l_1, l_2],$

$\varphi([-1, 0]) = [-p + r, -q + s] = [m_1, m_2],$

$\varphi([0, -1]) = [q + r, -p + s] = [n_1, n_2].$

We obtain the system of equations

$K \in a \Leftrightarrow h_1 : a_1(p + r) + a_2(q + s) + a_3 = 0,$

$L \in b \Leftrightarrow h_2 : b_1(-q + r) + b_2(p + s) + b_3 = 0,$

$M \in c \Leftrightarrow h_3 : c_1(-p + r) + c_2(-q + s) + c_3 = 0,$

$N \in d \Leftrightarrow h_4 : d_1(q + r) + d_2(-p + s) + d_3 = 0.$

Suppose that $a_1 = 0, a_2 = 1, a_3 = 0, c_3 = 0$ as in the previous case. We get the system of four *linear* equations with unknowns p, q, r, s. Then the elimination of q, r, s in the ideal $I = (h_1, h_2, h_3, h_4)$ gives the unknown p, etc. To compare this method of the solution with the previous one, we will add the polynomial $p+r-k_1$ to the ideal I and in the ideal $J = I \cup \{p+r-k_1\}$ eliminate p, q, r, s. In CoCoA we obtain

```
Use R::=Q[a[1..3]b[1..3]c[1..3]d[1..3]k[1]pqrs];
J:=Ideal(q+s,b[1](-q+r)+b[2](p+s)+b[3],c[1](-p+r)+c[2](-q+s),
d[1](q+r)+d[2](-p+s)+d[3],p+r-k[1]);
Elim(p..s,J);
```

the same result as above.

Remark 9.1. We solved the last problem in two different ways. It is obvious that the latter method was more effective and led to a faster solution (fewer unknowns, simpler equations, etc.).

Let us solve the example above by the classical method [59]. This solution is based on one theorem from equiform kinematics (see the book [145] by I. M. Yaglom). It says that if three points have straight trajectories in an equiform motion then all points have straight trajectories. By this theorem it suffices to construct two arbitrary squares X', Y', U', V' and X'', Y'', U'', V'' with only *three* vertices X', Y', U' and X'', Y'', U'' on given lines a, b, c. Then the remaining vertices V', V'' determine the line p which is a trajectory of a vertex N. The intersection of the lines p and d gives a vertex N of a square $KLMN$ (Fig. 9.2).

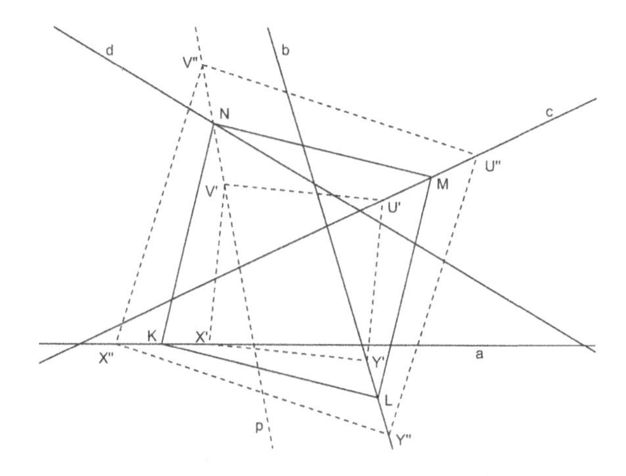

Fig. 9.2 Square $KLMN$ whose vertices lie on given lines a, b, c, d — classical method

Remark 9.2. In the previous problem we solved the system of four linear equations because four lines a, b, c, d were given. This problem can be generalized in the following way [59]:

Four algebraic curves p_1, p_2, p_3, p_4 in the plane are given. Construct a square $A_1A_2A_3A_4$ such that $A_i \in p_i$ for all $i = 1, \ldots, 4$.

If we take as curves p_1, p_2, p_3, p_4 conic sections then the resulting polynomial is of degree at most 16 (since $2^4 = 16$) and the problem becomes more complex.

Even in the geometry of a triangle, we would encounter non-elementary constructions. Let us look at the following problem:

Construct a right triangle ABC if the radius r of the inscribed circle and the length c of a hypotenuse are given.

We will use the coordinate-free method to describe the situation. Denote the side lengths of ABC by a, b, c. By the Pythagorean theorem, we have

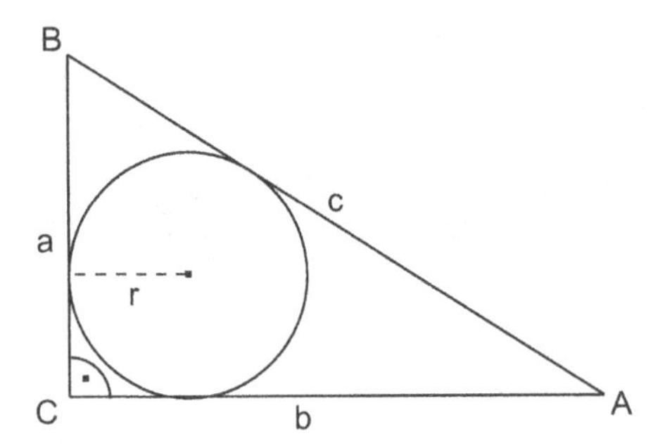

Fig. 9.3 Construction of a right triangle ABC from inradius r and hypotenuse c

(Fig. 9.3)

$h_1 : a^2 + b^2 - c^2 = 0.$

To get another condition we express the area S of ABC in two forms. We have either $S = 1/2ab$ or $S = r(a + b + c)/2$ and from it

$h_2 : ab = r(a + b + c).$

In order to express the side length a in terms of c and r, we eliminate a variable b in the ideal $I = (h_1, h_2)$. We enter

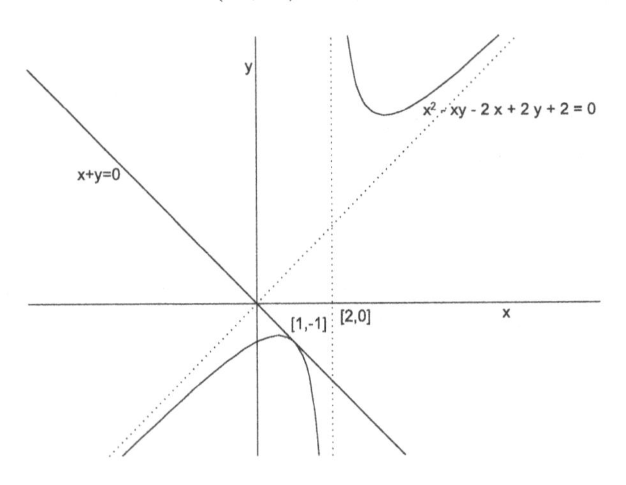

Fig. 9.4 The graph of the equation $x^3 - xy^2 - 2x^2 + 2y^2 + 2x + 2y = 0$

```
Use R::=Q[abcr];
```

```
I:=Ideal(a^2+b^2-c^2,ab-r(a+b+c));
Elim(b,I);
```

and obtain an algebraic equation of the 3rd degree in a:

$$a^3 - ac^2 - 2a^2r + 2c^2r + 2ar^2 + 2cr^2 = 0, \tag{9.2}$$

whose left side decomposes into linear and quadratic polynomials

$$(a + c)(a^2 - ac - 2ar + 2cr + 2r^2) = 0.$$

Hence, given c, r, we are able to construct by a ruler and compass the corresponding value a (see Fig. 9.4), where we put $r = 1$ and write x, y instead of a, c.

A classical solution is as follows. Let S be the incenter of a right triangle ABC (Fig. 9.5). For the angle ASB we get $\angle ASB = \pi - \alpha/2 - \beta/2 = \pi - \pi/4 = 3\pi/4$, since $\alpha + \beta = \pi/2$. First we construct a segment AB of the length c. Then a point S which is the intersection of a line p parallel to AB at the distance r and the arc k of a circle over AB from which we see AB under the angle $3\pi/4$. It is clear that we can get at most two solutions.

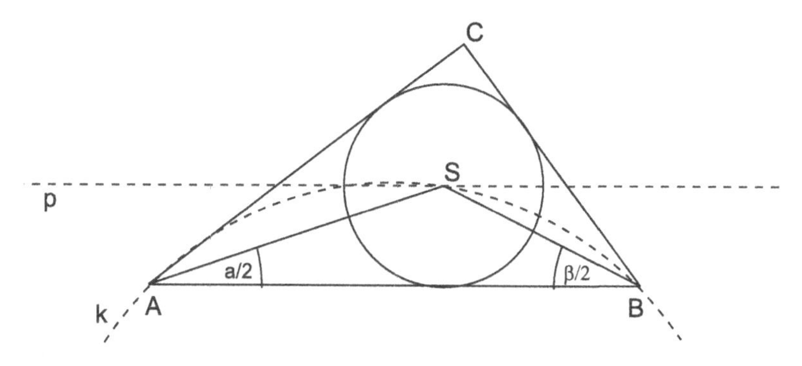

Fig. 9.5 Construction of a right triangle ABC from inradius r and hypotenuse c — classical solution

Let us compare the previous problem with the following one. We will explore:

Construct a right triangle ABC with the right angle at C if the side length b and the length u of angle bisector at B are given.

We see (Fig. 9.6) that from the Pythagorean theorem

$h_1 : a^2 + b^2 - c^2 = 0.$

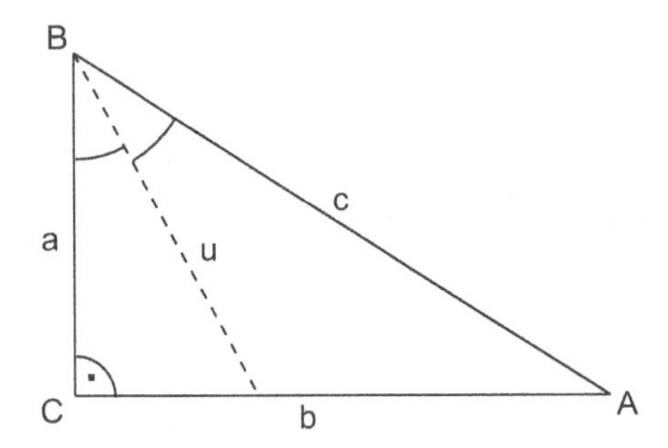

Fig. 9.6 Construction of a right triangle ABC from the side b and angle bisector u at the vertex B

The expression of the angle bisector u in terms of the side lengths a, b, c is as follows [4]:

$h_2 : u^2(a+c)^2 - ac((a+c)^2 - b^2) = 0.$

The elimination of c in the ideal $I = (h_1, h_2)$ gives

```
Use R::=Q[abcu];
I:=Ideal(a^2+b^2-c^2,u^2(a+c)^2-ac((a+c)^2-b^2));
Elim(c,I);
```

the algebraic equation of degree 6 in a:

$$4a^6 + 4a^4b^2 - 4a^4u^2 - 4a^2b^2u^2 + b^2u^4 = 0. \tag{9.3}$$

Denoting $a^2 = x$, $b^2 = y$, $u^2 = z$ then (9.3) can be written as a cubic equation in x (where y, z are given):

$$4x^3 + 4x^2y - 4x^2z - 4xyz + yz^2 = 0. \tag{9.4}$$

In the field of rational numbers, the left side of (9.4) can not be decomposed into the product of polynomials of lower degrees as factorization shows. Whence this problem can not be solved by ruler and compass [7]. Solving this problem by computer is possible — it suffices to find the roots x of the cubic equation (9.4) to any given y and z (see Fig. 9.7, where we put $z = 1$).

Remark 9.3. Non-elementary construction problems are often considered as uninteresting. The given examples show that the use of modern tools could change this approach.

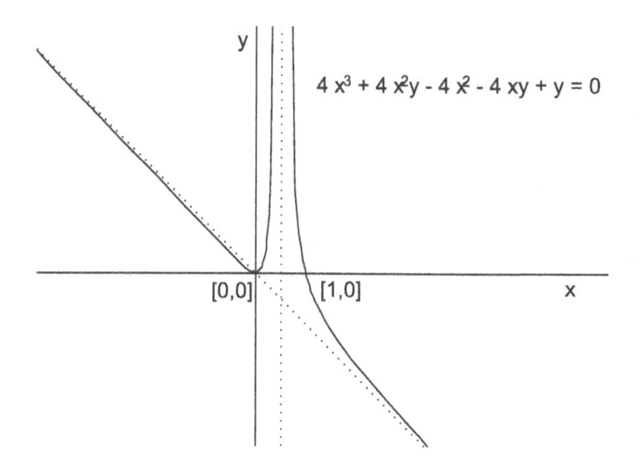

Fig. 9.7 The graph of the equation $4x^3 + 4x^2y - 4x^2 - 4xy + y = 0$

9.2 Loci of points of given properties

In this section we will investigate loci of points of given properties by the method of elimination of variables (see [136], Section 9). We are aware that this theme and its investigation by computer deserves more space (see also papers [13, 14], where a pair of computer programs Lugares and GDI specially devoted to computing loci of points in an automatic way using the same technique as in this book, is described). In the following example the author merely demonstrates the method of solving a typical problem. We have the following problem:

On the sides BC, CA, AB of a triangle ABC, the points D, E, F which divide the sides of ABC respectively in the same ratio k,

$$k = (BCD) = (CAE) = (ABF),$$

are given. Determine the locus of the circumcenters of triangles DEF when the value k varies.

Choose a Cartesian system of coordinates such that for the vertices of ABC $A = [0, 0]$, $B = [a, 0]$, $C = [b, c]$ (Fig. 9.8). Further denote $D = [d_1, d_2]$, $E = [e_1, e_2]$, $F = [f_1, f_2]$, where

$D \in BC \Leftrightarrow h_1 : cd_1 + (a - b)d_2 - ac = 0,$

$E \in CA \Leftrightarrow h_2 : ce_1 - be_2 = 0,$

$F \in AB \Leftrightarrow h_3 : f_2 = 0.$

For the ratios of points we have

$k = (BCD) \Leftrightarrow D - B = k(D - C) \Leftrightarrow$
$h_4 : d_1 - a - k(d_1 - b) = 0,\ h_5 : d_2 - k(d_2 - c) = 0,$

and analogously

$k = (CAE) \Leftrightarrow$
$h_6 : e_1 - b - ke_1 = 0,\ h_7 : e_2 - c - ke_2 = 0,$

$k = (ABF) \Leftrightarrow$
$h_8 : f_1 - k(f_1 - a) = 0.$

The circumcenter $S = [x, y]$ of a triangle DEF fulfils $S \in o_A \cap o_B$, where o_A and o_B are side bisectors of EF and FD, i.e.,

$S \in o_A \Leftrightarrow h_9 : (e_1 - f_1)x + e_2 y - \frac{1}{2}(e_1^2 - f_1^2 + e_2^2) = 0$

and

$S \in o_B \Leftrightarrow h_{10} : (d_1 - f_1)x + d_2 y - \frac{1}{2}(d_1^2 - f_1^2 + d_2^2) = 0.$

In the ideal $I = (h_1, h_2, \ldots, h_{10})$ we eliminate all variables up to a, b, c, x, y.

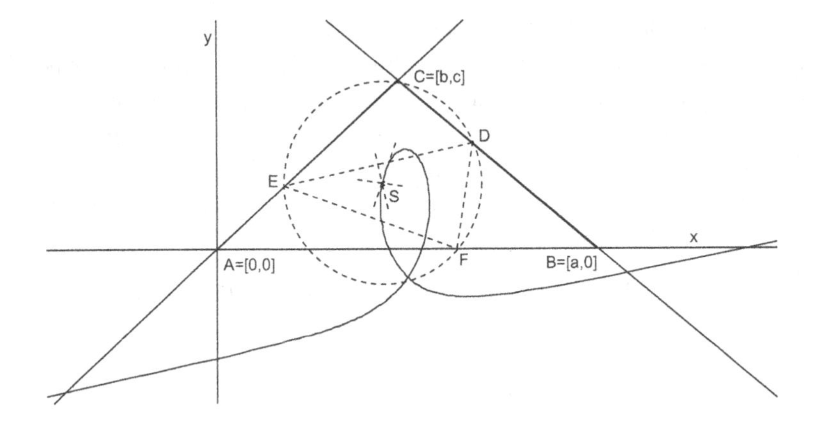

Fig. 9.8 Curve of 3rd degree as the locus of points of the given property

We enter

```
Use R::=Q[abckd[1..2]e[1..2]f[1..2]xy];
I:=Ideal(cd[1]+(a-b)d[2]-ac,ce[1]-be[2],f[2],d[1]-a-k(d[1]-b),
d[2]-k(d[2]-c),e[1]-b-ke[1],e[2]-c-ke[2],f[1]-k(f[1]-a),(e[1]-
f[1])x+e[2]y-(e[1]^2-f[1]^2+e[2]^2)/2,(d[1]-f[1])x+d[2]y-
```

```
(d[1]^2-f[1]^2+d[2]^2)/2);
Elim(k..f[2],I);
```

and CoCoA returns the polynomial of the third degree in x, y which leads to the equation

$$-8x^3(a-2b)(2a^2+ab-b^2-c^2)(a^2-ab+b^2)+24cx^2y(a^4-2a^3b+4ab^3-2b^4-a^2c^2+2abc^2-2b^2c^2)-24c^2xy^2(a-2b)(a^2+ab-b^2-c^2)+8c^3y^3(a^2+2ab-2b^2-2c^2)+4x^2(5a^6-9a^5b+a^3b^3-9ab^5+5b^6-5a^4c^2+7a^3bc^2+3a^2b^2c^2-14ab^3c^2+10b^4c^2+3a^2c^4-5abc^4+5b^2c^4)-4cxy(5a^5-8a^4b-a^3b^2+8a^2b^3+13ab^4-10b^5-9a^3c^2+8a^2bc^2+18ab^2c^2-20b^3c^2+5ac^4-10bc^4)+4c^2y^2(3a^4-2a^3b-5a^2b^2-4ab^3+5b^4-5a^2c^2-4abc^2+10b^2c^2+5c^4)-4x(2a^7-2a^6b-2a^5b^2-2a^2b^5-2ab^6+2b^7-3a^5c^2+2a^4bc^2+3a^3b^2c^2-4a^2b^3c^2-5ab^4c^2+6b^5c^2+3a^3c^4-2a^2bc^4-4ab^2c^4+6b^3c^4-ac^6+2bc^6)+4cy(a^6-a^5b-a^4b^2+a^3b^3+3a^2b^4+ab^5-2b^6-3a^4c^2+a^3bc^2+6a^2b^2c^2+2ab^3c^2-6b^4c^2+3a^2c^4+abc^4-6b^2c^4-2c^6)+a^8-2a^6b^2-a^4b^4-2a^2b^6+b^8-2a^6c^2+2a^4b^2c^2-6a^2b^4c^2+4b^6c^2+3a^4c^4-6a^2b^2c^4+6b^4c^4-2a^2c^6+4b^2c^6+c^8=0.$$

This curve is obviously a cubic curve (Fig. 9.8).

In a special case for $a=2, b=0, c=2$ we get

$$x^3+y^3-3x^2+xy-3y^2+2x+2y-1=0, \qquad (9.5)$$

which has a singular point at $[1,1]$ (Fig. 9.9).

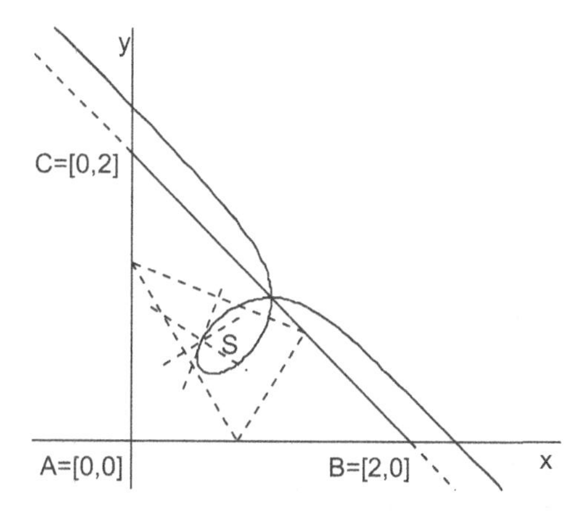

Fig. 9.9 Descartes' folium as the locus of points of a given property

Substituting $x'=x-1$ and $y'=y-1$ into (9.5) leads to

$$x'^3+x'y'+y'^3=0$$

which is the equation of the well-known *Descartes' folium.*

9.3 Theorem of Viviani

We shall explore the theorem of Viviani [10] and show how the validity of the theorem can be extended by computer to the whole plane.[1]

Viviani's theorem reads (Fig. 9.10):

Theorem 9.1 (Viviani). *Given an equilateral triangle ABC and an arbitrary point P inside ABC or on its perimeter. Then the sum of distances of P to the sides of ABC is constant which is equal to the length h of the altitude of ABC, i.e.,*

$$|PK| + |PL| + |PM| = h, \tag{9.6}$$

where K, L, M are the feet of perpendiculars from P to the sides BC, CA, AB respectively.

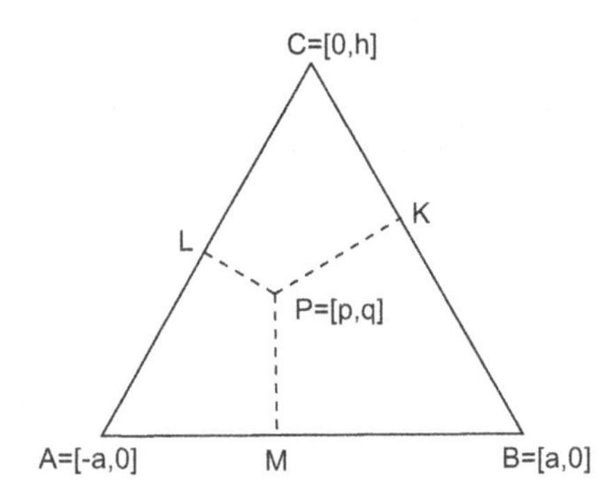

Fig. 9.10 Theorem of Viviani: Sum of distances of a point P to the sides of an equilateral triangle ABC is constant

In this case we first start with a classical proof. We show an elegant proof which is given in [10] (see also [79]).

Consider an equilateral triangle ABC and its shifted congruent image $A'B'C'$ such that the side $B'C'$ passes through the point P (Fig. 9.11). Then

$$h = |C'Q| + |PM| = |SP| + |PM| = |SL| + |LP| + |PM| =$$

[1]This example of discovery already appeared in the book [103] by T. Recio.

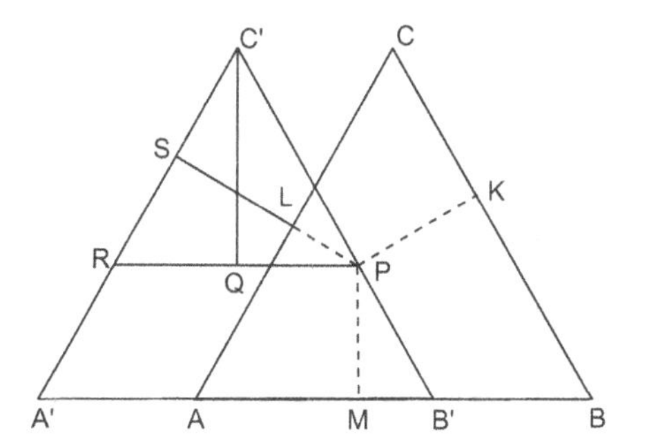

Fig. 9.11 Classical proof of theorem of Viviani: The sum of distances of a point P to the sides of an equilateral triangle ABC is constant

$$= |PK| + |PL| + |PM|.$$

Now we will prove Viviani's theorem by computer. In addition we will get even more — we generalize the theorem.

Choose a Cartesian system of coordinates so that $A = [-a, 0]$, $B = [a, 0]$, $C = [0, h]$, $P = [p, q]$, $K = [k_1, k_2]$, $L = [l_1, l_2]$, $M = [m_1, 0]$ (Fig. 9.10). We have:

$$PK \perp BC \Leftrightarrow h_1 : a(p - k_1) - h(q - k_2) = 0,$$

$$K \in BC \Leftrightarrow h_2 : ah - hk_1 - ak_2 = 0,$$

$$PL \perp AC \Leftrightarrow h_3 : a(p - l_1) + h(q - l_2) = 0,$$

$$L \in AC \Leftrightarrow h_4 : hl_1 - al_2 + ah = 0,$$

$$PM \perp AB \Leftrightarrow h_5 : a(p - m_1) = 0.$$

Denote $|PK| = u, |PL| = v, |PM| = w$. Then

$$|PK| = u \Leftrightarrow h_6 : (p - k_1)^2 + (q - k_2)^2 - u^2 = 0,$$

$$|PL| = v \Leftrightarrow h_7 : (p - l_1)^2 + (q - l_2)^2 - v^2 = 0,$$

$$|PM| = w \Leftrightarrow h_8 : (p - m_1)^2 + q^2 - w^2 = 0.$$

An equilateral triangle ABC with the side length $2a$ and height h fulfils

$$|AB| = |BC| = |CA| = 2a \Leftrightarrow h_9 : 3a^2 - h^2 = 0.$$

The conclusion c has the form:

$$c : u + v + w - h = 0.$$

Suppose that the point $P = [p, q]$ is *an arbitrary* point in the plane. Denoting $I = (h_1, h_2, \ldots, h_9)$ we will compute the normal form of 1 with respect to the Gröbner basis of the ideal $J = I \cup \{ct - 1\}$, where t is a slack variable. The result in CoCoA

```
Use R::=Q[apquvwhk[1..2]l[1..2]m[1..2]t];
J:=Ideal(a(p-k[1])-h(q-k[2]),ah-hk[1]-ak[2],a(p-l[1])+h(q-l[2]),
hl[1]-al[2]+ah,a(p-m[1]),(p-k[1])^2+(q-k[2])^2-u^2,(p-l[1])^2+
(q-l[2])^2-v^2,(p-m[1])^2+q^2-w^2,3a^2-h^2,(u+v+w-h)t-1);
NF(1,J);
```

is 1. Investigation of non-degeneracy conditions does not give any result as well. Hence, we will use the method B (see Section 2.5 for details), which is based on the extension of the conclusion variety. In the ideal J we eliminate all variables except u, v, w, h. We receive

```
Use R::=Q[uvwhapqk[1..2]l[1..2]m[1..2]t];
J:=Ideal(a(p-k[1])-h(q-k[2]),ah-hk[1]-ak[2],a(p-l[1])+h(q-l[2]),
hl[1]-al[2]+ah,a(p-m[1]),(p-k[1])^2+(q-k[2])^2-u^2,(p-l[1])^2+
(q-l[2])^2-v^2,(p-m[1])^2+q^2-w^2,3a^2-h^2,(u+v+w-h)t-1);
Elim(a..t,J);
```

the following polynomial

$h^2(u+v-w-h)(u-v+w+h)(u-v-w+h)(u-v+w-h)(u-v-w-h)(u+v+w+h)(u+v-w+h)$.

Suppose that $a \neq 0$ which implies $h \neq 0$. Then adding this polynomial to the conclusion polynomial $u + v + w - h$ (in fact we mean their product), we get the new conclusion polynomial c' which is of the form

$$c' : ((-u+v+w)^2-h^2)((u-v+w)^2-h^2)((u+v-w)^2-h^2)((u+v+w)^2-h^2). \tag{9.7}$$

Now the procedure will repeat. Instead of the conclusion c we consider *the extended* conclusion c'. Computation of the normal form of 1 in the ideal $K = I \cup \{as - 1, c't - 1\}$, where s is another slack variable,

```
Use R::=Q[apquvwhk[1..2]l[1..2]m[1..2]st];
K:=Ideal(a(p-k[1])-h(q-k[2]),ah-hk[1]-ak[2],a(p-l[1])+h(q-l[2]),
hl[1]-al[2]+ah,a(p-m[1]),(p-k[1])^2+(q-k[2])^2-u^2,(p-l[1])^2+
(q-l[2])^2-v^2,(p-m[1])^2+q^2-w^2,3a^2-h^2,as-1,((-u+v+w)^2-h^2)
((u-v+w)^2-h^2)((u+v-w)^2-h^2)((u+v+w)^2-h^2)t-1);
NF(1,K);
```

gives the result 0. We establish the following theorem:

Theorem 9.2. *Given an equilateral triangle ABC and an arbitrary point P in the plane of a triangle. Let u, v, w be the distances of the point P to the sides BC, CA, AB or their extensions respectively. Then*

$$((-u+v+w)^2-h^2)((u-v+w)^2-h^2)((u+v-w)^2-h^2)((u+v+w)^2-h^2)=0. \tag{9.8}$$

What is a geometric meaning of the condition (9.8)? The condition

$$(u+v+w)^2-h^2=0$$

which is equivalent to $|u+v+w| = |h|$ is fulfilled by the points P which are inside the triangle ABC or on its perimeter.
However, there are also other possibilities. Conditions

$$(-u+v+w)^2-h^2=0,\ (u-v+w)^2-h^2=0,\ (u+v-w)^2-h^2=0$$

express the fact that the "sum" of distances of a point P to the sides of the triangle ABC is constant also in the case when a point P is outside the triangle ABC. To express the given fact, it is advantageous to introduce the signed distance of a point to a straight line. The *signed distance* of a point P to a line AB is its distance to a line AB with the sign "+", if the sense of a motion on the perimeter of a triangle PAB from P to A and B is counter-clockwise, i.e., it is positive. In the opposite case the signed distance has the sign "−". We denote the signed distance of a point P to the line AB by $\|P, AB\|$. Hence, for instance $\|P, AB\| = -\|P, BA\|$. Using

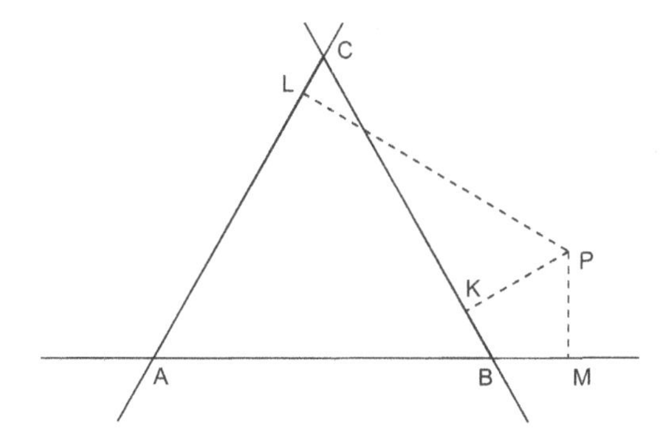

Fig. 9.12 $\|P,AB\| + \|P,BC\| + \|P,CA\| = w - u + v = h.$

signed distances we can rewrite Theorem 9.2 in the following simpler form (Fig. 9.12):

Theorem 9.3. *Given an equilateral triangle ABC and an arbitrary point P in the plane of a triangle. Then the sum of signed distances of a point P to the sides of a triangle obeys*

$$\|P, AB\| + \|P, BC\| + \|P, CA\| = \|A, BC\|. \tag{9.9}$$

Remark 9.4. The sum of the signed distances is often called the algebraic sum.

Let us give a classical proof of (9.9). For the area $\|ABC\|$ of an equilateral triangle ABC we have

$$\|ABC\| = \|PAB\| + \|PBC\| + \|PCA\|, \tag{9.10}$$

where $\|ABC\|$, $\|PAB\|$, $\|PBC\|$, $\|PCA\|$ are signed areas of triangles ABC, PAB, PBC, PCA (Fig. 9.13).

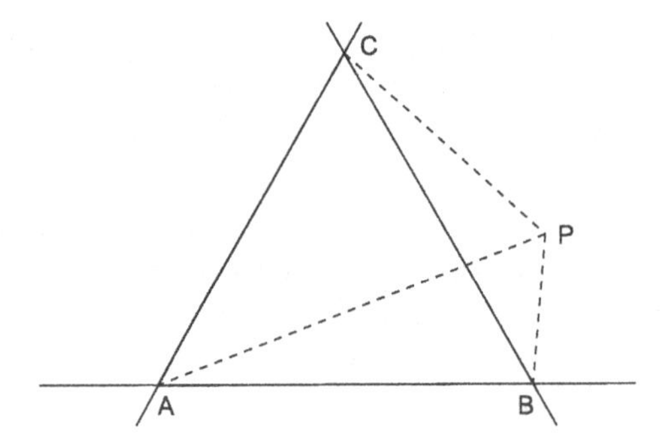

Fig. 9.13 Classical proof of the equality $\|P, AB\| + \|P, BC\| + \|P, CA\| = \|A, BC\|$.

Assume that the side length of ABC equals $2a$. Using the formula (9.10) we get

$$a\|A, BC\| = a\|P, AB\| + a\|P, BC\| + a\|P, CA\|.$$

From here (9.9) follows.

9.4 Gauss' line

The following problem is known as "the theorem on a complete quadrilateral" or "Gauss' line", see [5, 22, 126, 66, 134, 145]. We give this example to show that sometimes it is very easy to prove a theorem by computer but it is not easy to prove it synthetically.

Theorem 9.4 (Gauss' line). *The three midpoints of diagonals of a complete quadrilateral are collinear.*

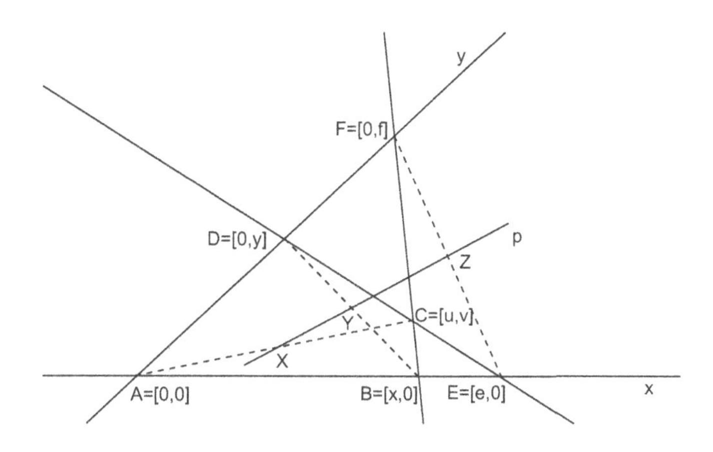

Fig. 9.14 Gauss' line — computer proof

A line p which contains all three midpoints is called the *Gauss' line.*
A computer proof is as follows. Choose an affine coordinate system such that $A = [0,0]$, $B = [x,0]$, $C = [u,v]$, $D = [0,y]$, $E = [e,0]$, $F = [0,f]$, $X = [p_1,p_2]$, $Y = [q_1,q_2]$, $Z = [r_1,r_2]$ (Fig. 9.14).

Then the following relations hold:

$E \in AB \cap CD \Leftrightarrow h_1 : ev + uy - ey = 0,$

$F \in AD \cap BC \Leftrightarrow h_2 : fu + vx - fx = 0,$

X is the midpoint of $AC \Leftrightarrow h_3 : 2p_1 - u = 0,\ h_4 : 2p_2 - v = 0,$

Y is the midpoint of $BD \Leftrightarrow h_5 : 2q_1 - x = 0,\ h_6 : 2q_2 - y = 0,$

Z is the midpoint of $EF \Leftrightarrow h_7 : 2r_1 - e = 0,\ h_8 : 2r_2 - f = 0.$

The conclusion z of the statement has the form

X, Y, Z are collinear $\Leftrightarrow z : p_1q_2 + q_1r_2 + r_1p_2 - r_1q_2 - r_2p_1 - p_2q_1 = 0.$

Computing the normal form of z with respect to the Gröbner basis of the ideal $I = (h_1, h_2, \ldots, h_8)$ (with some prescribed order of variables), we get in CoCoA

```
Use R::=Q[xyuvefp[1..2]q[1..2]r[1..2]];
```

```
I:=Ideal(ev+uy-ey,fu+vx-fx,2p[1]-u,2p[2]-v,2q[1]-x,2q[2]-y,
2r[1]-e,2r[2]-f);
NF(p[1]q[2]+q[1]r[2]+r[1]p[2]-r[1]q[2]-r[2]p[1]-p[2]q[1],I);
```

the result `NF(z,I)=0`. The statement is proved.

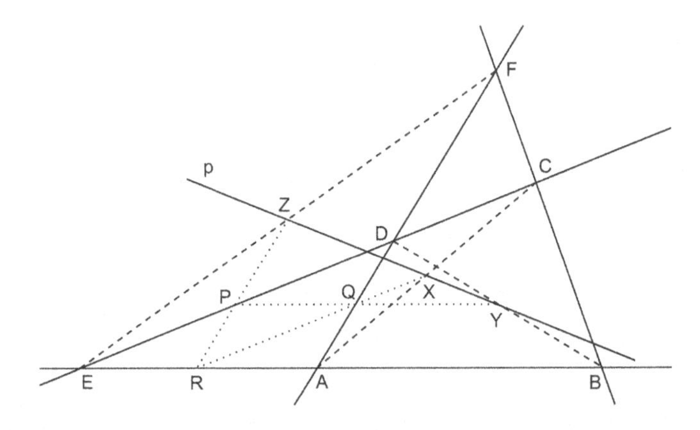

Fig. 9.15 Gauss' line — classical proof

Let us show a classical proof of the above statement. We will prove this theorem by means of Menelaus' theorem.

Consider a triangle ADE and let P, Q, R be the midpoints of its sides DE, AD and EA respectively (Fig. 9.15). We see that $PR \parallel AD$ and $Z \in PR$. Analogously $QR \parallel DE$ and $X \in QR$ and $QP \parallel EA$ and $Y \in QP$. In order to prove that the points X, Y, Z are collinear, we apply the theorem of Menelaus on the triangle PQR and the straight line p. We have to prove that

$$\frac{\|RX\|}{\|XQ\|} \cdot \frac{\|QY\|}{\|YP\|} \cdot \frac{\|PZ\|}{\|ZR\|} = -1. \tag{9.11}$$

We have

$$\frac{\|RX\|}{\|XQ\|} = \frac{\|EC\|}{\|CD\|}, \quad \frac{\|QY\|}{\|YP\|} = \frac{\|AB\|}{\|BE\|}, \quad \frac{\|PZ\|}{\|ZR\|} = \frac{\|DF\|}{\|FA\|} \tag{9.12}$$

and from here

$$\frac{\|RX\|}{\|XQ\|} \cdot \frac{\|QY\|}{\|YP\|} \cdot \frac{\|PZ\|}{\|ZR\|} = \frac{\|EC\|}{\|CD\|} \cdot \frac{\|AB\|}{\|BE\|} \cdot \frac{\|DF\|}{\|FA\|} = -1 \tag{9.13}$$

by the Menelaus' theorem applied to the triangle EDA, since the points C, B, F are collinear.

Remark 9.5. We proved Gauss' theorem automatically by computer and then classically. One may think that one of these methods is simpler than the other. In this case it seems that the computer approach is easier. But sometimes, the converse could happen.

9.5 Most common problems

The theory of automatic proving theorems together with solving many examples by this method was taught by the author at the University of South Bohemia for several years. This theme was not in common in the geometric seminar and the students were not inhabited to handle orderly specific tasks. Here are the most common problems which students encounter in automatic theorem proving [90]:

1) Introduction of a coordinate system (if necessary).

- There is an unsuitable introduction of the coordinate system.
- Which type of a coordinate system should we choose in order to "save" variables?
- Are we working in affine, Euclidean, projective or else geometry?
- Should we solve the problem by the coordinate or coordinate-free way?

2) Algebraization of a geometric situation.

- Complex or unsuitable descriptions of the geometric situation by algebraic equations results in many variables, large polynomials and high order of algebraic equations which lead to a big computational cost.
- Incorrect descriptions of a geometric situation might occur too (e.g. non-uniqueness).

3) A bad use of a computational method.

- Keeping the right order of all the stages of automatic proving.
- Computation of the normal form of an ideal.
- Elimination of variables, their ordering, which variables to eliminate — dependent, independent, semidependent.
- Searching for non-degeneracy (additional) conditions.

4) A geometric meaning of an algebraic equation.

- Understanding a geometric meaning of an algebraic equation — the major problem.

- Non-degeneracy conditions and their geometric interpretation.
- Formulation of a generalized statement.

5) Complex versus real numbers.[2]

- The difference between solutions in complex and real numbers.
- Getting formulas which are fulfiled only in complex numbers.

In the end let us give a survey of experiences we had obtained during a seminar and on the basis of seminar works:

- Students learnt that there exist another "modern" method to solve problems.

- The classical and computer approach are complementary.

- Students had many problems with computer solving — more than expected.

- Students rely more on computer instead of themselves.

- For problems with translation of a statement from algebra to geometry, e.g. to explain degeneracy conditions geometrically, students prefer algebraic formulation to geometric one, etc.

- Automatic theorem proving requires deep geometric knowledge.

- There exist problems which we are not able to solve classically; most noneuclidean constructions can be done by computer.

- The main principle — switching between geometry and algebra.

- Motivation to geometry and algebra.

- Learning English.

Open questions:

- To what extent to teach the theory of Gröbner bases computation, normal forms etc. in courses of mathematics (from "black box" to perfect understanding)?

- Is it necessary to include automatic theorem proving into mathematics teacher's curricula?

- Is it necessary to include automatic theorem proving (in reduced form)

[2]See [28], where automatic theorem proving in the real case is given in detail.

into secondary school curricula (the 15-18 age group)?

Conclusion

In this book, application of the theory of automatic proving, automatic deriving and automatic discovering theorems in a few geometric stories from elementary geometry were given. On many examples we could see the core of this theory which, especially in the last few years, due to the increasing power of computers and new efficient algorithms becomes meaningful.

The solutions of almost all examples are accompanied by classical ones. The author's effort was to stress that not only classical but also computer solutions contain ideas which show the beauty of both geometry and algebra. In any case we should not harbour any preference to either method as each has its own strengths and weaknesses. So it is not possible to say that one method is better than the other. Instead, both methods should complement each other.

There were also cases in which a classical method was not available or at least it was not known to the author. In such cases, we had to rely on some of the mentioned computer methods. It is a challenge for us to attempt to find a classical solution in these instances as well. One manifestation is the case of a generalization of the formula of Heron for a cyclic pentagon and another n-gons for higher n.

There also exist cases when we are able to solve a problem by a classical method while on the contrary a computer method fails. This is one more challenge — to develop the theory of automatic theorem proving so that it always terminates, i.e., that we are always able to say that a given statement is true or not.

Some problems chalked up a huge computational memory cost, making it impossible to solve more complex problems. An example is the generalization of the formula of Heron for a cyclic n-gons for $n \geq 6$, which we are not able to solve by the methods previously used before. One more challenge is to apply solution strategies which are not so time consuming. It would allow us to solve a wider range of problems.

A certain development will be achieved by enhancing the efficiency of computers, but the major and substantial progress in solving problems can be expected with better and more sophisticated algorithms and more ingenious ways of solutions to problems.

Besides the two mentioned basic methods of proving — classical and computer — the book should bring to the readers, as the author firmly hopes, joy and benefit through many known or less known results in geometry. Hence, even the reader who does not like computer methods should have benefited as well. In such a case it suffices to concentrate on classical solutions.[1]

---

[1]We forward the interested reader the following address, where he/she can find the complete reference collection on the topic
`http://www-calfor.lip6.fr/ wang/GRBib/Welcome.html`
At the address `http://webs.uvigo.es/adg2006/` one can link to the previous Conferences ADG (Automatic Deduction in Geometry) and find about their proceedings.

Bibliography

[1] Van Aubel, M. H.: Note concernant les centres des carres construits sur les côtes d'un polygone quelconque. *Nouv. Corresp. Math.* **4**: 40–44 (1878).

[2] Auric, A.: Question 3867. *Intermed. Math.* **18**: 122 (1911).

[3] Barlotti, A.: Una proprieta degli n-agoni che si ottengono transformando in una affinita un n-agono regolare. *Bol. Unione Math. Ital.* **3**: 96–98 (1955).

[4] Bartsch, H.-J.: *Taschenbuch mathematischer Formeln.* VEB Fachbuchverlag, Leipzig (1979).

[5] Berger, M.: *Geometry I, II.* Springer, Berlin Heidelberg (1987).

[6] Berkhan, G., Meyer, W. F.: Neuere Dreiecksgeometrie. In: *Encyklop. Math. Wiss. III AB.* **10**: 1173–1276 (1914).

[7] Bieberbach, L.: *Theorie der geometrischen Konstruktionen.* Leipzig (1952).

[8] Blaschke, W.: *Kreis und Kugel.* Walter de Gruyter & Co, Berlin (1956).

[9] Van der Blij, F.: Regular Polygons in Euclidean Space. *Lin. Algebra and Appl.* **226-228**: 345–352 (1995).

[10] Bogomolny, A.: `http://www.cut-the-knot.org/geometry.shtml` (2006).

[11] Bonnesen, T., Fenchel, W.: *Theorie der konvexen Körper.* Springer, Berlin (1934).

[12] Botana, F.: Automatic Determination of Algebraic Surfaces as Loci of Points. In: *Proceedings of ICCS 2003*, Lecture Notes in Computer Science **2567**, pp. 879–886. Springer, Berlin Heidelberg (2003).

[13] Botana, F., Valcare, J. L.: A software tool for the investigation of plane loci. *Mathematics and Computers in Simulation* **61(2)**: 139–152 (2003).

[14] Botana, F., Valcare, J. L.: Automatic determination of envelopes and other derived curves within a graphic environment. *Mathematics and Computers in Simulation* **67(1–2)**: 3–13 (2004).

[15] Bottema, O. *et al*: *Geometric inequalities.* Groningen (1969).

[16] Bottema, O.: Pentagons with equal sides and equal angles. *Geom. Dedicata* **2**: 189–191 (1973).

[17] Bottema, O.: *Hoofstukken uit de elementaire meetkunde.* Servire, Den Haag, (1944), Epsilon, Utrecht (1987).

[18] Buchberger, B.: Gröbner bases: an algorithmic method in polynomial ideal theory. In: *Multidimensional Systems Theory* (Bose, N.–K., ed.), pp. 184–

232. Reidel, Dordrecht (1985).
[19] Bulmer, M., Fearnley–Sander, D., Stokes, T.: The Kinds of Truth of Geometric Theorems. In: *Automated Deduction in Geometry* (Richter-Gebert, J., Wang, D. eds), Lecture Notes in Computer Science **2061**, pp. 129–142. Springer-Verlag, Berlin Heidelberg (2001).
[20] Bydžovský, B.: *Úvod do algebraické geometrie.* Jednota českých matematiků a fyziků, Praha (1948).
[21] Choi, M. D., Lam, T. Y., Reznick, B.: Even Symmetric Sextics. *Mathematische Zeitschrift* **195**: 559–580 (1987).
[22] Chou, S. C.: *Mechanical Geometry Theorem Proving.* D. Reidel Publishing Company, Dordrecht (1987).
[23] Conti, P., Traverso, C.: Algebraic and Semialgebraic Proofs: Methods and Paradoxes. In: *Automated Deduction in Geometry* (Richter-Gebert, J., Wang, D. eds), Lecture Notes in Computur Science **2061**, pp. 83–103. Springer-Verlag, Berlin Heidelberg (2001).
[24] Cox, D., Little, J., O'Shea, D.: *Ideals, Varieties, and Algorithms.* Second Edition, Springer, New York Berlin Heidelberg (1997).
[25] Cox, D., Little, J., O'Shea, D.: *Using Algebraic Geometry.* Second Edition, Springer, New York Berlin Heidelberg (1998).
[26] Coxeter, H. S. M., Greitzer, S. L: *Geometry revisited.* Toronto New York (1967).
[27] Dalzotto, G., Recio, T.: On protocols for the automated discovery of theorems in elementary geometry. Submitted to Journal of Automated Reasoning.
[28] Dolzmann, A., Sturm, T., Weispfenning, V.: A New Approach for Automatic Theorem Proving in Real Geometry. *Journal of Automated Reasoning* **21**: 357–380 (1998).
[29] Dörrie, B. H.: *Triumph der Mathematik.* Breslau (1933).
[30] Douglas, J.: Geometry of polygons in the complex plane. *J. Math. Phys. (MIT)* **19**: 93–130 (1940).
[31] Douglas, J.: On linear polygon transformations. *Bull. Amer. Math. Soc.* **46**: 551–560 (1940).
[32] Douglas, J.: A theorem on skew pentagons. *Scripta Math.* **25**: 5–9 (1960).
[33] Dunitz, J. D., Waser, J.: The Planarity of the Equilateral, Isogonal Pentagon. *Elem. der Math.* **27**: 25–32 (1972).
[34] Efremovitch, V. A., Iljjashenko, Ju. S.: Regular polygons. *Vestnik Mosk. Univ.* **17**: 18–24 (1962). In Russian.
[35] Euler, L.: *Novi commentarii academicae scientiarum Petropolitanae* **11**: 103–123 (1765,1767).
[36] Euler, L.: Geometrica et sphaerica quedam. *Mémoires de l'Académie des Sciences de St.* **5**: 96–114 (1812).
[37] Fan, K., Taussky, O., Todd, J.: Disctrete analogs of inequalities of Wirtinger. *Monatsh. Math.* **59**: 73–90 (1955).
[38] Fedorchuk, M., Pak, I.: Rigidity and polynomial invariants of convex polytopes. To appear in *Duke Math. J.* (2004), available at `http://www-mathmit.edu/~pak/research.html`

[39] Finney, R. L.: Dynamic proofs of Euclidean theorems. *Math. Mag.* **43**: 177–186 (1970).
[40] Finsler, P., Hadwiger, H.: Einige Relationen im Dreieck. *Comment. Math. Helv.* **10**: 316–326 (1937/38).
[41] Fischer, W.: Ein geometrischer Satz. *Arch. Math. Phys.* **40**: 460–462 (1863).
[42] Forder, H.: *The Calculus of Extension.* Chelsea, New York (1960).
[43] Gao, X. S., Wang, D. (eds.): *Mathematics Mechanization and Applications.* Academic Press, San Diego San Francisco New York Boston London Sydney Tokyo (2000).
[44] Gerber, L.: Napoleon's theorem and the parallelogram inequality for affine-regular polygons. *Amer. Math. Monthly* **87**: 644–648 (1980).
[45] Giering, O.: Affine and Projective Generalization of Wallace Lines. *J. Geom. and Graphics* **1**: 119–133 (1997).
[46] Gray, Stephen B.: Generalizing the Petr–Douglas–Neumann Theorem on N-gons. *Amer. Math. Monthly* **110**: 210–227 (2003).
[47] Grünbaum, B.: Polygons. In: *The Geometry of Metric and Linear Spaces*, Lecture Notes in Mathematics **490**, pp. 147–184. Springer, Berlin Heidelberg (1975).
[48] Grünbaum, B.: Cyclic ratio sums and products. *Crux Math.* **24**: 20–25 (1998).
[49] Grünbaum, B., Shephard, G. C.: Ceva, Menelaus, and the Area Principle. *Math. Magazine* **68**: 254–268 (1995).
[50] Grünbaum, B., Shephard, G. C.: Ceva, Menelaus, and Selftransversality. *Geometriae Dedicata* **65**: 179–192 (1997).
[51] Grünbaum, B., Shephard, G. C.: Some New Transversability Properties. *Geometriae Dedicata* **71**: 179–208 (1998).
[52] Guzmán, M: An Extension of the Wallace–Simson Theorem: Projecting in Arbitrary Directions. *Amer. Math. Monthly* **106**: 574–580 (1999).
[53] Hadwiger, H.: *Altes und Neues über konvexe Körper.* Birkhäuser Verlag, Basel und Stuttgart (1955).
[54] Hardy, G. H., Littlewood, J. E., Pólya, G.: *Inequalities.* Second Edition, Cambridge University Press (1952).
[55] Hilbert, D., Cohn–Vossen, S.: *Anschauliche Geometrie.* Berlin (1932).
[56] Hora, J., Pech, P.: Using computer to discover some theorems in geometry. *Acta Acad. Paed. Agriensis* **29**: 67–75 (2002).
[57] Irminger, H.: Zu einem Satz über räumliche Fünfecke. *Elem. der Math.* **25**: 135–136 (1970).
[58] Kapur, D.: A Refutational Approach to Geometry Theorem Proving. *Artificial Intelligence Journal* **37**: 61–93 (1988).
[59] Karger, A.: Classical Geometry and Computers. *Journal for Geometry and Graphics* **2**: 7–15 (1998).
[60] Kárteszi, F.: Contributo al pentagono equilatero ed isogonale. *Ann. Univ. Sci. Budapest, Roland Eötvös Sect.* **16**: 63–64 (1973).
[61] Kiepert, L.: Solution de question 864. *Nou. Ann. Math.* **8**: 40–42 (1869).
[62] Kline, J. S., Velleman, D. J.: Yet another proof of Routh's theorem. *Crux. Math.* **21**: 37–40 (1995).

[63] Kočandrlová, M.: The Isoperimetric Inequality for a Pentagon in E^3 and Its Generalization in E^n. *Čas. pro pěst. mat.* **107**: 167–174 (1982).
[64] Kočandrlová, M.: Isoperimetrische Ungleichung für geschlossene Raumsechsecke. *Čas. pro pěst. mat.* **108**: 248–257 (1983).
[65] Koepf, W.: Gröbner Bases and Triangles. *International Journal of Computer Algebra in Mathematics Education* **4**: 371–386 (1997).
[66] Konforovič, A. G.: *Významné matematické úlohy.* SPN, Praha (1989).
[67] Kowalewski, G.: *Einführung in die Determinantentheorie.* Veit & Comp., Leipzig (1909).
[68] Kutzler, B.: Careful algebraic translations of geometry theorems. In: *Proceedings ISSAC '89*, Portland, pp. 254–263. ACM Press, New York (1989).
[69] Lawrence, J.: K-equilateral $(2k+1)$-gons span only even-dimensional spaces. In: *The Geometry of Metric and Linear Spaces*, Lecture Notes in Mathematics **490**, pp. 185–186. Springer, Berlin Heidelberg (1975).
[70] Lüssy, W., Trost, E.: Zu einem Satz über räumliche Fünfecke. *Elem. der Math.* **25**: 82–83 (1970).
[71] Malay, F. M., Robbins, D. P., Roskies, J.: On the areas of cyclic and semicyclic polygons. `arXiv:math.GM/0407300` (2004).
[72] Martini, H.: Neuere Ergebnisse der Elementargeometrie. In: *Geometrie und Ihre Anwendungen* (Giering, O., Hoschek, J., eds), pp. 9–42. Carl Hanser Verlag, München und Wien (1994).
[73] Martini, H.: On the theorem of Napoleon and related topics. *Math. Semesterber.* **43**: 47–64 (1996).
[74] Merriell, D.: Further remarks on concentric polygons. *Amer. Math. Monthly* **72**: 960–965 (1965).
[75] Mitrinovic, D. S., Pecaric, J. E., Volenec, V.: *Recent Advances in Geometric Inequalities.* Kluwer Acad. Publ., Dordrecht, Boston, London (1989).
[76] Míšek, B.: O $(n+1)$-úhelníku v E^n s maximálním objemem konvexního obalu. *Čas. pro pěst. mat.* **84**: 93–103 (1959).
[77] Naas, J., Schmid, H. L.: *Mathematisches Wörterbuch.* Berlin Leipzig (1967).
[78] Nádeník, Z.: Náměty k středoškolské geometrii. *Pokroky matematiky, fyziky a astronomie* **19**: (1974).
[79] Nelsen, R.: *Proofs Without Words.* MAA (1993).
[80] Nelsen, R.: *Proofs Without Words II.* MAA (2000).
[81] Neumann, B. H.: Some remarks on polygons. *J. London Math. Soc.* **16**: 230–245 (1941).
[82] Novoselov, S. I.: *Special'nyj kurs trigonometrii.* Nauka, Moskva (1953).
[83] Nudel'man, A. A.: Izoperimetričeskije zadači dlja vypuklych oboloček lomanych i krivych v mnogomernych prostranstvach. *Mat. Sbornik* **96**: 294–313 (1975).
[84] Osserman, R.: Bonnesen-Style Isoperimetric Inequalities. *Amer. Math. Monthly* **86**: 1–29 (1979).
[85] Pak, I.: The area of cyclic polygons: Recent progress on Robbins' Conjectures. `arXiv:math.MG/0408104` (2004)
[86] Parrilo, P. A.: Structured Semidefinite Programs and Semialgebraic Geo-

metry Methods in Robustness and Optimization. *PhD. thesis.* California Institute of Technology, Pasadena, California (2000).
[87] Pech, P.: Inequality between sides and diagonals of a space n-gon and its integral analog. *Čas. pro pěst. mat.* **115**: 343–350 (1990).
[88] Pech, P.: The harmonic analysis of polygons and Napoleon's theorem. *J. Geometry and Graphics* **5**: 13–22 (2001).
[89] Pech, P.: On Simson–Wallace Theorem and Its Generalizations. *J. Geometry and Graphics* **9**: 141–153 (2005).
[90] Pech, P.: Classical versus computer algebra methods in elementary geometry. *Int. Journal of Technology in Mathematics Education* **12** 137–148 (2005).
[91] Pech, P.: A Harmonic Analysis of Polygons. *Slov. J. Geom. and Graphics* **2**: 17–36 (2005).
[92] Pech, P.: Computations of the Area and Radius of Cyclic Polygons Given by the Lengths of Sides. In: *Automated Deduction in Geometry* (Hong, H., Wang, D. eds.), Lecture Notes in Artificial Intelligence **3763**, pp. 44–58. Springer, New York Heidelberg (2006).
[93] Pech, P.: On the Need of Radical Ideals in Automatic Proving: A Theorem about Regular Polygons. In: *Automated Deduction in Geometry 2006* (Botana, F., Recio, T. eds.), Lecture Notes in Artificial Intelligence, Springer, New York Heidelberg. To appear.
[94] Pedoe, D.: An Inequality for Two Triangles. *Proc. Cambridge Philos. Soc.* **38**: 397–398 (1943).
[95] Pedoe, D.: *Geometry.* Dover Publ., New York (1988).
[96] Peschl, E.: *Analytische Geometrie und lineare Algebra.* Bibliographisches Institut AG, Mannheim (1961).
[97] Petr, K.: O jedné větě pro mnohoúhelníky rovinné. *Čas. pro pěst. mat. a fyz.* **34**: 166–172 (1905).
[98] Petr, K.: Ein Satz über Vielecke. *Arch. Math. Physik* **13**: 29–31 (1908).
[99] Prasolov, V. V., Šarygin, I. F.: *Zadači po stereometriji.* Nauka, Moskva (1989).
[100] Rashid, M. A., Ajibade, A. O.: Two conditions for a quadrilateral to be cyclic expressed in terms of the lengths of its sides. *Int. J. Math. Educ. Sci. Techn.* **34**: 739–742 (2003).
[101] Recio, T., Sterk, H., Vélez, M. P.: Project 1. Automatic Geometry Theorem Proving. In: *Some Tapas of Computer Algebra* (Cohen, A., Cuipers, H., Sterk, H., eds.), Algorithms and Computations in Mathematics, Vol. **4**, pp. 276–296. Springer, New York Heidelberg (1998).
[102] Recio, T., Vélez, M. P.: Automatic Discovery of Theorems in Elementary Geometry. *J. Automat. Reason.* **12**: 1–22 (1998).
[103] Recio, T.: *Cálculo simbólico y geométrico.* Editorial Síntesis, Madrid (1998).
[104] Recio, T., Botana, F.: Where the Truth Lies (in Automatic Theorem Proving in Elementary Geometry). In: *Proceedings ICCSA (International Conference on Computational Science and its Applications)*, Lecture Notes in Computer Science **3044**, pp. 761–771. Springer-Verlag, Berlin Heidelberg (2004).

[105] Rédey, L., Nagy, B. Sz.: Eine Verallgemeinerung der Inhaltsformel von Heron. *Publ. Math. Debrecen* **1**: 42–50 (1949).

[106] Reznick, B.: Some concrete aspects of Hilbert's 17th Problem. *Contemp. Math.* **253**: 257–272 (2000).

[107] Riesinger, R.: On Wallace Loci from the Projective Point of View. *J. Geom. and Graphics* **8**: 201–213 (2004).

[108] Rigby, J. F.: A concentrated dose of old-fashioned geometry. *The mathematical Gazette* **57**: 297–298 (1973).

[109] Rigby, J. F.: Napoleon Revisited. *J. Geom.* **33**: 129–146 (1988).

[110] Rigby, J. F.: Napoleon, Escher and Tessellations. *Structural Topology* **17**: 17–23 (1991).

[111] Roanes, E. M., Roanes–Lozano, M.: Automatic Determination of Geometric Loci. 3D-Extension of Simson–Steiner Theorem. In: *AISC 2000*, (Campbell J. A., Roanes–Lozano, E. eds), Lecture Notes in Artificial Intelligence **1930**, pp. 157–173. Springer, New York Heidelberg (2001).

[112] Robbins, D. P.: Areas of polygons inscribed in a circle. *Discrete Comput. Geom.* **12**: 223–236 (1994).

[113] Routh, E. J.: *A Treatise on Analytical Statics, with Numerous Examples.* Vol. I, 2nd ed., Cambridge University Press, London (1896).

[114] Sadov, S.: Sadov's Cubic Analog of Ptolemy's Theorem. `http://www.math.rutgers.edu/$\sim$zeilberg/mamarim/mamarimhtml/` `\\sadov.html` (2004).

[115] Schoenberg, I. J.: The Finite Fourier Series and Elementary Geometry. *Amer. Math. Monthly* **57**: 390–404 (1950).

[116] Schoenberg, I. J.: An isoperimetric inequality for closed curves convex in even-dimensional Euclidean spaces. *Acta Math.* **91**: 143–164 (1954).

[117] Schoenberg, I. J.: The Harmonic Analysis of Skew Polygons as a Source of Outdoor Sculptures. In: *The Coxeter Festschrift, Geometric Vein*, pp. 165–176. Springer, Berlin (1982).

[118] Schreiber, P.: On the Existence and Constructibility of Inscribed Polygons. *Beiträge zur Algebra und Geometrie* **34**: 195–199 (1993).

[119] Scriba, Ch. J.: Wie kommt "Napoleon Satz" zu seinem Namen? *Hist. Math.* **8**: 458–459 (1980).

[120] Sekanina, M. *et al*: *Geometrie I.* SPN, Praha (1986).

[121] Shephard, G. C.: Euler's Triangle Theorem. *Crux Math.* **25**: 148–153 (1999).

[122] Shephard, G. C.: Cyclic sums for polygons. *Math. Magazine* **72**: 126–132 (1999).

[123] Shephard, G. C.: The Compleat Ceva. *Math. Gazette* **83**: 1–8 (1999).

[124] Simon, M.: *Über die Entwicklung der Elementar-Geometrie im XIX. Jahrhundert.* Teubner, Leipzig (1906).

[125] Škljarskij, D. O., Čencov, N. N., Jaglom, I. M.: *Geometričeskije neravenstva i zadači na maximum i minimum.* Nauka, Moskva (1970). In Russian.

[126] Škljarskij, D. O., Čencov, N. N., Jaglom, I. M.: *Izbrannyje zadači i teoremy elementarnoj matematiky, časť 2 Geometrija (planimetrija).* Moskva (1952). In Russian.

[127] Šmakal, S.: Eine Bemerkung zu einem Satz über räumliche Fünfecke. *Elem. der Math.* **27**: 62–63 (1972).

[128] Staudt, Ch. R.: Über die Inhalte der Polygone und Polyeder. *Journal für die reine und angewandte Mathematik* **24**: 252–256 (1842).

[129] Svrtan, D., Veljan, D., Volenec, V.: Geometry of pentagons: from Gauss to Robbins. Available at `http://arxiv.org/abs/math/0403503` (2004).

[130] Tarski, A.: The completness of elementary algebra and geometry. Institut Blaise Pascal, Paris (1967), iv + 55pp. Reprinted from a paper which was prepared to printing in 1940 in publishing house Hermann & Cie but was not edited due to the World War the 2nd.

[131] Thébault, V.: Solution to Problem 169. *National Math. Mag.* **12**: 192–194 (1937–38).

[132] Van der Waerden, B. L.: Ein Satz über räumliche Fünfecke. *Elem. der Math.* **25**: 73–96 (1970).

[133] Van der Waerden, B. L.: Nachtrag zu "Ein Satz über räumliche Fünfecke". *Elem. der Math.* **27**: 63 (1972).

[134] Wang, D.: Gröbner Bases Applied to Geometric Theorem Proving and Discovering. In: *Gröbner Bases and Applications* (Buchberger, B., Winkler, F. eds.), Lecture Notes of Computer Algebra, pp. 281–301. Cambridge Univ. Press, Cambridge (1998).

[135] Wang, D.: *Elimination Methods.* Springer-Verlag, Wien New York (2001).

[136] Wang, D.: *Elimination Practice. Software Tools and Applications.* Imperial College Press, London, (2004).

[137] Waser, J.: Ph.D. thesis, Calif. Institute of Techn., Pasadena (1944).

[138] Weiss, G.: Written communication.

[139] Weitzenböck, R.: *Math. Zeitschrift* **5**: 137–146 (1919).

[140] Wetzel, J. E.: Converses to Napoleon's theorem. *Amer. Math. Monthly* **99**: 339–351 (1992).

[141] Wieleitner, H.: *Geschichte der Mathematik, II. Teil* (1911–1921), Russian translation (1958), 2. edition (1966).

[142] Winkler, F: Gröbner Bases in Geometry Theorem Proving and Simplest Degeneracy Conditions. *Mathematica Pannonica* **1**: 15–32 (1990).

[143] Wu, W: On the Decision Problem and the Mechanization of Theorem Proving in Elementary Geometry. *Scientia Sinica* **21**: 157–179 (1978).

[144] Wu, W.: *Mechanical Theorem Proving in Geometries.* Texts and Monographs in Symbolic Computation, Springer, (1994).

[145] Yaglom, I. M.: *Geometric transformations.* Vol. II. Random House (1968).

[146] Yang, L., Zhang, J.: A Practical Program of Automated Proving for a Class of Geometric Inequalities. In: *Automated Deduction in Geometry* (Richter-Gebert, J., Wang, D. eds.), Lecture Notes in Computer Science **2061**, pp. 41–57. Springer-Verlag, Berlin Heidelberg (2001).

[147] Zhang, Xin-Min: Bonnesen-style inequalities and pseudo-perimeters for polygons. *Journal of Geometry* **60**: 188–201 (1997).

Index